NEW FOUNDATIONS OF
QUANTUM MECHANICS

NEW FOUNDATIONS OF
QUANTUM MECHANICS

BY

ALFRED LANDÉ

Emeritus Professor of Physics
Ohio State University

CAMBRIDGE
AT THE UNIVERSITY PRESS
1965

The text of this book rests on articles published in *The Physical Review* (U.S.A.); *American Journal of Physics* (U.S.A.); *Philosophy of Science Journal* (U.S.A.); *British Journal for the Philosophy of Science* (Great Britain); *Endeavour* (Great Britain); *Mind* (Great Britain); *Journal de Physique* (France); *Physikal. Zeitschrift* (Germany); *Naturwissenschaften* (Germany); *Philosophia Naturalis* (Germany); *Current Science* (India); *Dialectica* (Switzerland). It is a further development of the author's *Foundations of Quantum Theory, a Study in Continuity and Symmetry*, Yale and Oxford University Presses, 1955, and *From Dualism to Unity in Quantum Mechanics*, Cambridge University Press, 1960. About twenty pages from the latter are taken over into the present book.

CAMBRIDGE
UNIVERSITY PRESS

University Printing House, Cambridge CB2 8BS, United Kingdom

Cambridge University Press is part of the University of Cambridge.

It furthers the University's mission by disseminating knowledge in the pursuit of education, learning and research at the highest international levels of excellence.

www.cambridge.org
Information on this title: www.cambridge.org/9781107593541

© Cambridge University Press 1965

First published 1965
First paperback edition 2015

A catalogue record for this publication is available from the British Library

ISBN 978-1-107-59354-1 Paperback

CONTENTS

CONTENTS

CONTENTS

CONTENTS

PREFACE

This book is a much enlarged and consolidated version of the author's *Foundations of Quantum Theory, a Study in Continuity and Symmetry* (Yale University Press, 1955) and *From Dualism to Unity in Quantum Physics* (Cambridge University Press, 1960). Its aim is the solution of a problem known to an earlier generation as 'The Quantum Riddle', that is, the task of deriving the amazing laws of quantum mechanics from a non-quantal basis without *ad hoc* assumptions. Non-quantal is not to mean classical deterministic, however. In the contrary, our starting-point is probabilistic from the beginning. By applying the same general postulates of symmetry and invariance, which are known from deterministic mechanics, also to regulate the structure of a probabilistic theory, one arrives at the 'new' quantum mechanics which was established in 1926 after more than two decades of trial and error. The Quantum Riddle is thereby solved by the method of theoretical physics.

But scientists, bent on research and development of new territory, are seldom swayed by methodological considerations alone to revise their once accepted approach. The innovation of this book is to point out that there are serious faults in the purely *physical* arguments which have led to the current dualistic doctrine according to which diffraction and other wavelike phenomena of matter force us to accept two contradictory pictures, together with an elaborate subjectivist interpretation of atomic events, instead of one unitary reality. The faulty physics consists in ignoring an important element of the quantum mechanics of particles, namely the rule for the quantized exchange of linear momentum (Duane's third quantum rule, 1923) established in perfect analogy to the quantum rule for the energy (Planck) and for the angular momentum (Sommerfeld-Wilson). The third quantum rule yields indeed a complete explanation of all the wavelike phenomena of matter, including diffraction and coherence, without using the fantastic hypothesis of particles occasionally transforming themselves into waves or, what is the same, 'manifesting' themselves as though they did. It is a grave error of omission when physicists in their extended discussions about the interpretation of quantum theory never mention the third rule of quantum mechanics, the Missing Link between wave-like appearances and particle reality, entailing an agonizing reappraisal of the antiquated yet ever repeated dogma of duality. The latter is a thorn in the sides of both physics and natural philosophy which we are trying to extract.

There are, then, two distinct keys struck in this study. First, a major one of deducing the quantum laws from general non-quantal postulates of symmetry and invariance imposed on the probability structure of mechanical events. It is a new key, as far as I know, and as far as the literature resounds with assurances that quantum mechanics can *not* be so deduced, that it rather rests on its own sovereign principles. Second, a minor key, struck by many others before, of criticizing the current tenet according to which the knowledge of observers, subjective pictures, refined language, and renunciation of objective reality are inherent in the theory.

If the amazing formalism of quantum mechanics can be understood as a necessity under simple non-quantal assumptions, this encourages a new approach to the *teaching* of the theory in a systematic organic fashion, as a *quantum design*, rather than as 'the strange story of the quantum' which follows the erratic ways and byways of history. Instead of assimilating the once useful tale of two pictures, the student may learn from the beginning that quantum mechanics is neither excessively abstract and symbolic, nor does it deal with the knowledge or absence of knowledge of observers; it rather connects data of instruments, as does every other branch of theoretical physics.   A. L.

NOTE

Superior figures in the text thus (^{1})
refer the reader to References on
pages 166–8.

CHAPTER I

DUALISM VERSUS QUANTUM MECHANICS

'We are to admit no more causes of natural things than such as are both true and sufficient to explain their appearances.'

NEWTON

1. *Description, interpretation, explanation*

The principal aim of this book, apart from criticizing the current quantum philosophy of the Schools, is to give an *explanation* of the laws of the quantum theory on a simple and elementary non-quantal basis. This program may look like a futile excursion into thin air to a generation of physicists who have been told again and again that the quantum theory can *not* be reduced to anything more fundamental, that its laws are basic and inexorable features of Nature—an opinion which seemed to be confirmed by the many abortive efforts, during the first decade of this century, to detect a flaw in classical statistical mechanics so as to make room for Planck's discrete energy levels, efforts which proved entirely futile. Indeed, if one wishes to derive the quantum laws from a more elementary basis, one must start from probabilistic considerations right away, forget about classical mechanics, and be glad if the latter comes out as a statistical approximation.

But then, what does it mean *explaining* the quantum laws, and physical laws in general? For example take the magnetic properties of atoms. They became a topic of intense interest through Zeeman's magnetic splitting of spectral lines, afflicted with certain unexpected features, however. The Anomalous Zeeman Effect was deciphered in 1922 by the author through reduction of the observed atomic frequencies to magnetic energy levels, revealing various gyro-magnetic or

1

g-factors of unknown origin, involving half-quantum numbers for angular momenta in a vector model. The quantum number $\frac{1}{2}$ was later *interpreted* by Goudsmit and Uhlenbeck as peculiar to a spinning electron. Finally, the spin was *explained* by Dirac through a combination of relativity theory and quantum principles. The aim of this book is to go still farther and to explain the quantum principles themselves, that is to show them to be consequences of still more elementary principles known from pre-quantal physics.

For another example of description *versus* explanation take the theory of relativity. A student of physics learning that travelling rods contract, traveling clocks slow down, and mass defects correspond to surplus energies according to $E = mc^2$ would hardly be satisfied with being told that these are basic and inherent qualities of matter which can no longer be explained. But Einstein's derivation of the relativistic phenomena from general postulates of symmetry and invariance applied to a combination of mechanics and optics gave the desired explanation. Why then should we accept the quantum laws such as $E = h\nu$ as irreducible and inherent qualities of the physical world, or even less convincing as 'consequences of the dualism of the wave picture and the particle picture'? (For these quotations, refer to Chapter VIII.) The belief in a dual nature of matter and of light as *fundamental* has indeed blocked the search for a better understanding of the quantum theory for more than three decades, promoting the view that the student 'after learning the [mathematical] tricks of the trade' will finally 'understand that there is nothing to be understood' about the 'why' of the quantum laws. The present book is to challenge this purely descriptive approach; it intends to show that the perplexing wavelike interference of probabilities and the quantum rules, $E = h\nu$, etc., can be *explained* as necessary consequences of general postulates of symmetry and invariance imposed on the structure of a probability metric—as Einstein's $E = mc^2$ has been derived from similar postulates applied to a combination of mechanics and optics in which the velocity of light is the measure of space and time.

The aim of the first chapter is to show that electron diffraction and related wave like phenomena can be explained by pure mechanical particle action without wave interference.

2. *Diffraction through crystals*

The chief exhibit in defense of a dual nature of matter was, and to many physicists still is, the famous matter diffraction experiment. A homogeneous ray of electric charge, clearly consisting of individually countable electrons with concentrated charges, is sent toward a crystal from which the ray is reflected. As required by the conservation laws of energy and momentum, the angle of reflection equals that of incidence. The curious fact is, however, that reflection takes place only at certain discrete angles of incidence so as to produce discrete points or lines rather than a broad band of reflected intensity on a film, a most amazing sight and one of the showpieces of atomic physics. What could be the cause of this selective reflection?

It turned out that one can *calculate* the discrete angles of reflection by imagining that each single electron, arriving with momentum p, spreads out near the crystal into a broad wave aggregate of wave length $\lambda = h/p$ (h is Planck's constant) covering the whole crystal, that these 'matter waves' are reflected simultaneously from the parallel lattice planes spaced at distance l, whereupon wave interference, according to the Huyghens' principle of superposition, yields strongly reflected intensities only in those angular directions which belong to path differences of one, or two, or n wave lengths. Having done their duty, the waves re-form into particles again, as evident from the statistical build-up of the reflected pattern by individual impacts on a film. Interference maxima occur in those directions θ which satisfy the rule (Fig. 1a):

$$2l \sin \theta = n\lambda \text{ (Bragg)} \tag{1}$$

in which the left-hand side represents the path difference of two wave trains reflected from two subsequent lattice planes.

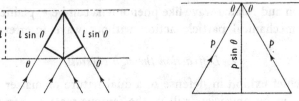

Fig. 1 a, b

There is no objection to the supernatural or, as others see it, broadminded(?) transmutation hypothesis, provided that it is regarded as a first probing step into unknown territory, pending further clarification. But instead of trying to clarify the mystery on physical grounds, physicists are committed to regard double manifestation as an unshakable truth, and evade the problem of the *real* constitution of matter, either waves or particles, by a sophisticated skepticism toward the idea of physical reality. Maybe everything from trees to neutrinos and galaxies is a mental construction fashioned into a Gestalt out of chaotic sense impressions; we know this anyway. But in science, as in ordinary life, the term 'real' is meant in no other sense than that a cloud really consists of individual water droplets, although it may appear as a continuous gray substance. And an apparently smooth image on a film in reality consists, as closer inspection shows, of many individual grains blackened one after another in a statistical manner. Nor does it help to remove the dilemma of diffraction if one replaces the supernatural forth-and-back transmutation by a mysterious 'dual manifestation' (a purely linguistic device), or if one speaks of the wave 'picture' in reaction to the crystal, and of the particle 'picture' away from the crystal. The task of physical theory is not inventing figures of speech but understanding atomic phenomena by a consistent theory representing one physical reality. If electrons are once condensed within a range of 10^{-12} centimeter or less, how can they ever spread out—or, as some would say, manifest themselves as though spreading out—over a billion times that range? Although it is generally believed that, without double manifestation, the diffraction of matter and other coherence phenomena would be impossible to explain, the impossible was achieved as early as 1923 by the American physicist W. Duane, one of the almost forgotten pioneers of modern quantum mechanics.

3. *Mechanics of diffraction without waves*

According to Duane [1], the incident matter particles do not spread out as continuous matter waves, or manifest themselves as though they did. It is the crystal with its parallel lattice planes which is already spread out in space and reacts as one rigid mechanical body to the incident particles under the conservation laws of mechanics with the following restriction (Duane's quantum rule for linear momentum):

A body periodic in space with linear periodicity of length L is thereby entitled to change its linear momentum p parallel to L in amounts $\Delta p = h/L$.

The foremost example of a space-periodic body is a crystal. If l is the mutual distance within a set of parallel lattice planes, the crystal contains not only the periodicity $L = l$ but also higher periodicities $L = \frac{1}{2}l, \frac{1}{3}l$, and so forth; hence, according to Duane's quantum rule, it can change its momentum perpendicular to the lattice planes in amounts

$$\Delta p = h/l, \ 2h/l, \ 3h/l, \ \ldots \ nh/l. \tag{2}$$

When these selective momenta are transferred from the crystal to the incident particles, the latter are deflected into exactly the same discrete directions which can also be *calculated* according to the Laue-Bragg wave interference theory. Yet this result is now obtained in a purely mechanical particle fashion, without double manifestation or any other 'causal anomaly' (Reichenbach's term) being involved. Indeed, when a particle of momentum p is incident and reflected at the same angle θ, as required by the conservation laws, its momentum component parallel to L changes as much as $2p \cdot \sin \theta$. The quantum rule above then yields the relation (Fig. 1b, p. 4).

$$2p \cdot \sin \theta = nh/l \text{ (Duane)} \qquad (3)$$

which is identical with Bragg's wave rule (1) by virtue of translating the momentum p into a wave length, $\lambda = h/p$, near the crystal and retranslating into the particle momentum p again after the reflection—only that *Duane's rule yields the same observed deflection directly without appealing to a wave interlude.* Therefore, the idea of a dualistic change from matter particles to waves is a quite unnecessary, if not fantastic, invention though it may sometimes be helpful to use: '*Se non è vero, è ben trovato*'.

Duane's quantum rule for the change of linear momentum by a body of periodicity L is by no means invented *ad hoc* (in contrast to the duality doctrine). It rather is the legitimate counterpart to the two familiar quantum rules for the energy E and for the angular momentum p_ϕ, namely Planck's rule, $\Delta E = h/T$, and Sommerfeld-Wilson's rule, $\Delta p_\phi = h/2\pi$, for bodies periodic in time with period T (harmonic oscillator) and periodic with respect to rotation through 360 degs. $= 2\pi$ (every body). If the system contains several inherent time periods $T_1, T_2, \ldots$ (an atom, as revealed by its spectrum) then it can change its energy in any one amount $\Delta E = h/T_n$ (Bohr's frequency condition). If the body has higher angular periodicities, if it is a regular polygon of periodicity $2\pi/n$, then it can change its angular momentum in amounts $\Delta p_\phi = nh/2\pi$. All three quantum rules are of course very perplexing. The common opinion is that we must accept them at face value, as fundamental and irreducible quantum principles. Yet, as we intend to show later in this study, they can be *explained* as being necessary consequences of simple and general postulates of symmetry and invariance, without any quantum ingredients to start with. In the present Introduction we take the quantum rules for granted.

Strangely enough, however, whereas the first two quantum rules are recognized as the most important instruments of the quantum theorist, the third rule for the linear momentum, $\Delta p = h/L$, is hardly ever mentioned. Yet, by way of Duane's theory of selective reflection of particles from a crystal, it deals a fatal blow to the doctrine of a wave-particle duality of matter, to the dogma that we must be content with the 'extravaganza' (Schrödinger's term) of two alternating pictures instead of one physical reality. Thus, when we are told authoritatively:[2] 'Quantum mechanics has discovered that the same physical object, for example an electron, appears in two seemingly exclusive forms', then we must reply that, in the contrary, quantum mechanics has discovered that even such wavelike phenomena as matter diffraction through crystals can be understood in a consistent unitary way as produced by matter particles alone obeying the conservation laws of mechanics under special restrictions, known as quantum rules, which apply to bodies containing periodicities in time and space. Electrons always behave as particles; *they never misbehave as waves.* It is one thing to instruct the adept that he may obtain results for the statistical behaviour of many particles by using a wavelike mathematical method, for example the Schrödinger equation, and then re-interpret the wave intensity as a statistical distribution density of particles. It is quite another thing to proclaim *urbi et orbi* that matter has a dual physical nature, that it appears as two pictures, sometimes as waves, another time as particles. Although this doctrine has attracted much attention, it has become a fallacy by virtue of Duane's third quantum rule which certainly ought to be granted equal time with the two other quantum rules.

One has retorted here: 'Duane regards the crystal as a mechanism acting through its periodic space components, whereas others assume that electrons occasionally spread out. What difference does it make?' There is a large difference in scientific outlook indeed. According to duality, a small particle miraculously becomes periodic and spreads out in space.

According to quantum mechanics and Duane, the crystal is there already; it is periodic and extended in space without extra 'manifestations'. Duality asks us to believe that a *thing* transforms itself into another thing, or acts as though it did. This is magic and will be denoted as such throughout this book. Quantum mechanics requires us only to accept an apparently strange (quantum) *action*. And this action can be explained as being quite natural (Chapters VI and VII) on elementary grounds. Duality is an extremely implausible ideology admittedly introduced *ad hoc* in order to cope with an apparent paradox, whereas Duane's quantum rule is legitimate physics. And yet, the dogma of a supernatural double manifestation has become so solidly entrenched today that one recalls Sidney Hook's general remark: 'The difference between science and religion is that the former wishes to get rid of mysteries whereas the latter worships them.'

It must be emphasized here that there is a large difference between the rules of quantum mechanics, $\Delta E = h/T$, $\Delta p = h/L$, and $\Delta p_\phi = h/2\pi$, for the energy and momentum exchange of physical bodies periodic in time and space, as against the formulas $E = h\nu$ and $p = h/\lambda$, which translate mechanical data of single particles into the picture of waves. Either the rules for energy and momentum, or the translation of particles into matter waves represent legitimate physics. I still trust that quantum theorists, when seriously challenged to choose between the two sets of rules, will decide in favour of the physical rules for energy and momentum exchange, and will recognize the translation rules as ideological crutches. This is not to dispute the tremendous influence exercised by de Broglie's original idea of matter waves and its value for heuristic purposes. But if Newton is right in the motto of this chapter, then the unitary theory of matter particles, that is quantum mechanics, is preferable to the uneconomical and *entirely superfluous* hypothesis of dualism, and is preferable also to a unitary theory of waves which is possible only at the cost of very great complications (Section 36).

Dualism stops at the observation that matter sometimes *looks like* particles and at other times *looks like* waves; so let us not be one-sided and take this 'looking like' as a principle. However, there are people who are less broadminded and would like to know what is behind the two subjective pictures or models. To them the answer is contained in the third quantum rule of mechanics which is as legitimate as the other two.

The reason that Duane's third quantum rule was not immediately recognized as a way out of the duality paradox was historical: Duane proposed his statistical particle theory of diffraction for X-rays in support of the photon theory of light. At his time (1923) diffraction of electrons was not yet discovered. Moreover, the quantum rule $\Delta p = h/L$ for bodies of periodicity L in space seemed artificial and conceived *ad hoc*. Since 1926, however, all three quantum rules have been victorious. There is no reason for ignoring the third rule alone, and relying instead on a principle of duality which gives a more or less fitting name to a dilemma rather than solving it in a scientific manner. Today, after endless repetition, a dual nature of matter may seem as obvious and indisputable to the experts as the immobility of the Earth seemed to Galileo's learned colleagues who refused to look through his telescope because it might make them dizzy. Yet Duane did as much to dissolve the phantom of duality as Gregor Mendel did to dispose of the age-old myth of inheritance through continuous blood mixture in favor of the statistical gene-particle theory. But Duane will probably have to wait as long as the Austrian abbot before his statistical interpretation and explanation of the wavelike diffraction phenomena are recognized as decisive steps toward modern quantum mechanics, as the Missing Link between wavelike appearances and particle reality.

4. *The two-slit experiment*

Let us next consider the much-discussed experiment of matter diffraction through a screen with one or two slits. A single slit yields a diffuse intensity distribution centered around the direct line of incidence of the matter ray. If now another adjacent slit is opened, various formerly bright places become dark. How can one explain that the *opening* of a second slit *blocks* particles from places at which they formerly arrived, without temporary transformation of the particles into waves and their interference? This is indeed the crucial question, discussed in scores of books, magazines, symposiums, and keynote speeches. The scientific solution of the problem is contained in a well-neglected investigation by Ehrenfest and Epstein[3] representing a further development of Duane's theory. An incident particle does not react to this or that individual slit (or bar in a reciprocal experiment), nor does it react to this or that lattice point of a crystal. A slit is more than a mere Nothing; it is a Nothing with something around. And the latter is as important as the former. A screen with one slit has periodic space components of various lengths L composing its geometrical shape; a screen with two slits has a different set of L's. And since the several components L give rise to impulse transfers $\Delta p = h/L$ respectively, the two cases of one and two slits yield different deflected angles with different intensities. The diffraction patterns produced statistically by deflected particles agree in both cases with those obtainable *also* via translation into waves and retranslation into particles. Yet the hypothesis of two alternating models to account for the observations proves expendable again.

Therefore, to the eternal question 'through which of the two slits does the electron pass' (supposing it does not perform the miracle of spreading out over both slits), the answer is: 'For its contribution to the diffraction pattern it does not make any difference where exactly the deflection takes place.' The electron changes its momentum in reaction to the harmonic components of the matter distribution of the screen with two slits as a whole; and the deflected electron may not even be identical with the incident one. All that matters is conservation of charge and of total momentum during the reaction between electron and diffractor.

It certainly is remarkable that the result of the quantum mechanical reaction of particles can also be obtained in the roundabout way of a wave calculation. However, this success of the wave theory is only a partial one; one needs discrete particles to account for the statistical build-up of the diffraction pattern. The quantum mechanics of particles without wave manifestation explains all features at the same time. If Schrödinger has justly characterized the duality notion as 'an extravaganza dictated by despair over a grave crisis', the crisis has been solved in a simple scientific manner without magic transmutations or manifestations. Yet in the numerous discussions of the matter diffraction phenomena allegedly forcing us to accept duality, Duane's solution by way of the third quantum rule is kept like a closely guarded secret.†

When an oscillator of frequency ν gives out an energy quantum $\Delta E = h\nu$ the latter can be taken over only by another system capable of the same frequency, that is another oscillator, or an atom containing $\nu = \Delta E/h$ in its discrete spectrum, or by the surrounding radiation which sustains a continuity of frequencies with ν among them, or also a free particle capable of any ΔE. In short, energy conservation during quantized energy exchange is conditioned by *frequency resonance*. In contrast to Schrödinger's effort to *replace* the 'energy jumps' by frequency resonance, both go hand in hand. A corresponding consideration holds for momentum exchange between space-periodic bodies. To imagine the quantized energy and momentum exchange as occurring at sharply localized instances and places, as jumps, is going beyond both experience and theory.

The quantum rules $\Delta E = h\nu$, etc., are usually accepted

† Max Born, in a letter to the author, writes: 'Duane's 1923 paper on the particle theory of X-ray diffraction was well appreciated at the time of its publication.' It then becomes even more of a riddle why its significance was overlooked when the diffraction of matter was discovered a few years later and then was ignored for thirty-odd years. I have searched the writings of Bohr, Born, de Broglie, Dirac, Einstein, Heisenberg, Pauli, Schrödinger, and many others for any hint of recognition of the third quantum rule as relevant to the alleged dilemma of matter diffraction and duality. There is no trace of it.

as fundamental, as not further explainable. It will be seen later in this book, however, that the quantum rules are direct consequences of the requirement that the probabilistic theory of mechanical processes (= quantum mechanics) be invariant with respect to shifts of zero points in space, time, momentum space, and energy.

5. *Coherence*

A frequent argument in favor of the wave theory is the apparent impossibility of explaining the phenomena of *coherence* without using interference of waves. We shall see, however, that the quantum theory of particles, with the help of the third quantum rule, yields exactly the same observed results. An (almost) monochromatic ray may be split in two parts which afterwards are recombined, either without path difference, yielding (almost) sharp maxima of intensity when sent through a diffractor, or with gradually fading-out maxima when the path difference is more and more increased, that is when the common stretch is shortened. Let us discuss the experiment first from the wave point of view.

The mean life span δt_0 of the source determines the coherence length δL_0 along the wave train. It is well known that a wave aggregate of length δL_0, when subjected to a spectral analysis by a diffractor, turns out to cover a range of wave lengths λ and corresponding 'wave numbers' $\kappa = 1/\lambda$ of width $\delta\kappa_0 = 1/\delta L_0$. If the ray is split in two, and the parts are recombined without path difference, the combination again has spectral width $\delta\kappa_0$. A spectral analyzer (diffracting instrument, crystal, or grating) will thus display maxima slightly diffused signified by the width $\delta\kappa_0$. Next let the parts be recombined with a path difference so that the common path is reduced from δL_0 to a smaller length δL. This restriction in space involves a larger spectral width, $\delta\kappa = 1/\delta L$, hence a more diffuse diffraction pattern. When δL approaches *zero* then the spectral width $\delta\kappa$ of the recombination becomes *infinite*, and no distinct maxima will occur.

Now translate the formalism of the *hypothetical* wave consideration into the *physics* of matter particle mechanics and remember that the third quantum rule explains the sharp intensity maxima at certain angles of diffraction in the ideal case of a 'monochromatic' ray of particles, all having exactly the same momentum p. However, when the particles come from a source of mean life δt_0 belonging to a path length δL_0, this physical limitation in space involves, according to the Heisenberg uncertainty theorem, a statistical spread of the particle momenta over a range $\delta p_0 \sim h/\delta L_0$, leading to a dispersion of the angles θ of reflected intensity obtained from a diffractor (crystal), according to Duane's rule (3). The same broadening of the maxima occurs when the particle ray is split in two, and the parts are recombined without path difference, the common path still having the length δL_0. However, when the common path is deliberately reduced to the shorter length δL, this limitation in space is connected, according to the uncertainty rule, with a broadening of the statistical range of the momentum values to $\delta p \sim h/\delta L$ which gives rise, according to Duane's theory, to a further diffusion of the diffracted maxima. When δL becomes *zero*, then δp becomes *infinite*, and the (vanishing) common path gives no discrete diffraction maxima. The result is independent of the absolute intensity of the matter ray. It holds also when the particles follow one another at large distances. All this is pure particle theory, also known as quantum mechanics. 'Coherence', shown by the presence of discrete maxima, depends on the length of the common path δL. But the same holds in wave theory where it is irrelevant for the coherence issue whether the common length δL contains many waves λ, or only a fraction of a wave length or none at all. This is the decisive reason for the equivalence of the two theories of coherence.

Similar considerations answer the question how particle mechanics, and the third quantum rule in particular, can explain the *traveling* of the whole system of maxima of various orders when the path difference of two superimposed parts of a ray is changed. Two slits in a screen produce a pattern of maxima. When the distance between the slits is changed, the path difference of the two rays combined in any point of the film is changed, too. According to the wave theory, this will shift the position of the maxima because of wave interference. The same result is obtained from particle mechanics. Indeed, changing the slit distance changes the periodicities L contained in the space structure of the screen with slits and thereby changes the sidewise momenta, $\Delta p = h/L$, given to the incident particles. The result agrees with that of the wave interference theory, according to Epstein-Ehrenfest's extension of Duane's theory. To repeat therefore: It is emphatically wrong to say that, according to *modern* physics, an electron is a 'wavicle', half wave and half particle.

6. Reflection from a wall

A doctrine as formidable as dualism ought to rely not only on the one or two cases for which it was designed (and where it proves dispensible) but elsewhere, too. Consider for example the reflection of particles from a plane wall. Since an ordinary wall does not have a periodic structure, or contains periodic space components L of any length whatsoever, it can give out any momentum $\Delta p = h/L$ without selectivity. Hence reflection takes place under the laws of energy and momentum conservation at all angles without restriction. A dualist, however, who does not wish to abandon his cause under fire, would have to maintain that the incident matter ray first and last manifests itself as the picture of particles, and on the wall manifests itself as the picture of waves producing secondary waves (Huyghens' principle) which by their interference are responsible for the angular reflection law. But why resort to the fantastic idea of double manifestation when the unitary mechanics of particles is sufficient before, during, and after the reflection?

A similar situation is found in case of *refraction*. Snell's law may be explained according to Newton's emission theory by letting particles of light acquire a larger momentum p on the boundary, according to Huyghens' undulatory theory by letting light waves acquire a smaller wave length λ. Here we have the root of the old dualistic controversy for light, bridged by the equation $p = h/\lambda$. Later the problem arose also for matter. In the same year 1923, in which Duane explained wavelike diffraction phenomena by his third quantum rule for particles, de Broglie proposed matter waves for the explanation of quantized energy and angular momentum values of atoms. Today a decision on the grounds of accumulated evidence has been reached by the unitary quantum mechanics of matter particles and the unitary quantum mechanics of light waves. According to the theory of radiation in reaction to electric matter (Dirac, Fermi, Heitler), there are neither continuous matter waves filling the whole space, nor discontinuous 'photons' dashing about, but there are *matter particles* in quantized reaction to *light waves*. Thus one may say that for matter Newton's emission theory, and for light Huyghens' undulatory theory has proved victorious. Neither for matter nor for light is there any need of having it both ways at the same time.

The question has been asked, most persistently by H. V. Stopes-Roe in several letters to *Nature*, as to *how*, that is by what mechanical action, a crystal of periodicity L can act as a whole mechanical unit comprising many periods L spread over a considerable space, so as to change its momentum according to the quantum rule $\Delta p = h/L$. Is it not more reasonable to assume that each incident electron spreads out like a wave over many lengths L, as the dualists have it? To my mind, the latter hypothesis gives far more grounds for uneasiness than the action of a system as a whole entity. He who asks how a whole periodic crystal can react under the quantum rule $\Delta p = h/L$ at one place, being 'conscious' of its overall periodicity, may as well ask *how* a body as a whole can change its angular momentum under the rule $\Delta p_\phi = h/2\pi$, being conscious of its periodicity under an angular rotation of

2π, and how an oscillator of time period T can change its energy in amounts $\Delta E = h/T$. (The difference is that the two latter cases are well known and accepted, whereas the former case seems to be unknown.) In short, how can a periodic body display concerted action under the quantum rules? It can act like this only in the case of instantaneous communication—which certainly does not hold in the relativistic realm, so that the quantum rules are applicable here only with reservations. How can the Pauli exclusion principle apply to two electrons, one here and the other on Sirius? Still I think that the quantum rules of mechanics are vastly more realistic than the idea of a transmutation of matter from particles to waves and back to particles again. Whether the latter is eulogized as a mental, subjective, unreal double manifestation 'as-if' makes little difference.†

7. *Discourse on Method*

In his popular booklet, *The Strange Story of the Quantum*, Banesh Hoffmann([4]) writes about the duality problem: 'If space and time are not the fundamental stuff of the universe but are merely . . . statistical effects of crowds of more fundamental entities lying deeper down, it is no longer strange that these fundamental entities should exhibit such ill-matched properties as those of wave and particle. There may, after all, be some innate logic in the quantum paradoxes.' I fully agree with the last sentence and with the request that ill-matched properties ought to be explained physically, rather than be 'manifested' away. But I do not think that one needs to go to new entities beyond space and time. The three quantum rules of mechanics are sufficient to lead from the strange story to a natural history of the quantum. Yet, in spite of the work of Duane and Born, the talk about duality has survived as a favorite topic for popular consumption. It is the starting-

† In a similar fashion, a periodic field component of radiation spread out in space and time transmits an energy quantum $h\nu$ to another body (electron oscillator, atom) not by local Maxwell-Lorentz action but *as a whole* under quantal probability and resonance rules during the interaction.

point of a new theory of knowledge in which pictures, even contradictory ones, have taken the place of the traditional search for physical reality, disregarding the fact that through the statistical interpretation and observation, and through the third quantum rule, a unitary aspect of matter has been recovered. Even more, the quantum rules themselves, in contrast to the 'pictures', can be *explained*, that is derived from simple and general propositions of a non-quantal character, as shown in the later chapters of this book. Dualism, on the other hand, is but a descriptive word, and a misleading one, too. It has led to the pseudo-philosophical opinion that questions about the real constitution of matter must not be asked any more because all things, from stars to trees and from particles to waves, are mere constructs of our minds, are pictures, so that the word 'real' ought to be put in quotation marks when referring to 'things'. Very well! But then one ought to put the waves of quantum theory in double quotation marks, as representing statistical distributions of particles, as pictures within the particle picture. My suggestion is to indulge in such luxury only in philosophical discourse, if ever, and to speak in the laboratory and in ordinary life without restraint of stars, trees, and atomic particles as real ('kickable': Dr. Johnson) things, and to put only the unkickable ψ-waves in quotation marks because they do not refer to things but are condensed probability tables. It clearly is a malfunction of elementary logic to establish with pride (and prejudice) a duality and complementarity of a material object and a probability table of events, as though the two were on an even conceptual level. According to the same logic, patients of undulating fever would be dualistic because, on the one hand, they are individuals, on the other hand, the probability of their being found in a hospital has ups and downs like a wave.

In a way, objectivity is inter-subjectivity. Therefore, as long as some people saw particles and others saw waves, it was a justified compromise to promote the view that matter does not *have* a definite objective constitution, and that everything is

17

a mental picture anyway. But if all people, as far as they are aware of the third quantum rule, see particles even in case of matter diffraction, this makes particles the only real constituents of matter, as the chemists found out a century and a half ago.

But once one has accepted *la grande illusion* of duality—that electrons sometimes misbehave as though filling the whole space (instead of correctly saying that the statistical distribution of many electrons sometimes yields a wavelike pattern)—there is no end to further misunderstanding. If a particle is not always a particle it means that 'we must not think in objects any more', that 'there is a collapse of the category of substance', and 'we must adapt our logic to the new situation' (von Weizsäcker's *World View of Physics*). I for my part refuse to adapt my logic, stop thinking in objects, and deplore the collapse of the category of substance every time I see electrons forming a periodic intensity pattern after having passed a periodic crystal. The periodic intensity is not due to a transmutation or manifestation in wave form of the electronic particles; its physical cause is the periodicity of the crystal itself, as Duane discovered. He thus explained a physical experiment in a physical manner instead of invoking supernatural powers.

It is true that the diffraction experiment together with the Schrödinger wave theory *seemed* to give a stunning blow to the idea of a corpuscular structure of matter notwithstanding the innumerable confirmations of the latter. In this quandary the talk that matter just has a dual nature was a welcome consolation. Schrödinger himself, however, asked for consistency and tried to develop a *unitary wave theory* of matter in which particles were high crests of waves. On the other hand, Born applied to matter what Duane had proposed for light: he interpreted the Schrödinger waves in a statistical corpuscular fashion. If Born's unitary particle interpretation is right, which hardly anyone doubts, then it is illogical to construe a dualistic opposition between a *thing*, a particle, as against *one* of its many qualities, a symbolic curve representing expectations to find a particle located here or there, even if the curve looks wavelike. This contrast of incontrastibles has enveloped the quantum theorists in self-contradictions and in erosion of logic. To mention a few examples, it has led to seven different interpretations, listed in Section 42, of the most common symbol of quantum theory, the letter ψ, as well as to a gratuitous re-interpretation of Heisenberg's uncertainty of prediction into an indeterminacy of existence by a literal translation of wave qualities into particle language (Section 38). If certain theories in economics held today are, according to the late President Kennedy, 'like old records, long playing, left over from the middle thirties', the same may be said of the orthodox theory of duality.

Some theorists take the stand that we should never try to form a graven image of microphysical things and events anyway, and should be content with having a perfect symbolic method for calculating observed phenomena. This formalistic approach seemed justified at a time when Heisenberg's and Born's intuition revised the older Bohr theory of electronic orbits by deriving observed data by means of abstract algebraic operations in a calculus of Hermitian matrices which nobody could *understand* in its physical significance. The same occurred again when Dirac introduced a still more enigmatic symbolic method which turned out, to his own surprise, to yield the observed half-integral spin and the magnetic moment of the electron. Still, these daring mathematical adventures were followed by the query what it all was to *mean* physically, until the very question about the meaning of those abstract methods was declared meaningless. Instead of encouraging the search for a plausible explanation of the quantum method, one tells us that it is enough to have an economical way of calculating phenomena. And if there are two conflicting descriptions, one of waves and another of particles, this is supposed to bring out the epistemological lesson that we should not bother about any 'real' world and had better accept the renun-

19

ciation to two pictures. No wonder that Sir James Jeans([5]), after having studied Bohr and Heisenberg, arrived at the triumphant conclusion: 'Nature consists of waves which are of the general quality of waves of knowledge, or of absence of knowledge, in our own minds,' thereby confusing a wavelike symbolic curve depicting the expectation odds for possible events with what Nature consists of. I rather agree with Schrödinger([6]) when he wrote: 'A widely accepted school of thought maintains that an objective picture of reality, in the traditional sense of the term, cannot exist at all. . . . I consider myself as one . . . who looks upon this view as a philosophical extravaganza born of despair in the face of a grave crisis.' Einstein, too, was always opposed to the as-if-picture view in physics. Following him, I regard electrons, positrons, etc., still as the same *things* although displaying some unexpected probabilistic *actions*, rather than as new things called 'wavicles'. Particles are still particles by virtue of their fixed charge, rest mass, and condensation within a small volume; their wavelike *appearances* are perfectly accounted for by the quantum rules of mechanical action.

Some theorists feel that they have developed an unfailing instinct for telling when and where matter changes its manifestation. Being aware of Duane's rule I do not believe in different manifestations at all. The particle theory suffices in all cases, whereas the continuous wave picture applies only to half the cases; it fails to account for the statistical build-up of the diffraction pattern. Nobody disputes that the grains of the film are triggered by electronic particles. I strongly deny, however, that the latter have to pass through a wave interregnum, real or imagined. Duality owes its stubborn life only to a historical accident: forgetting to transfer Duane's particle mechanics of X-ray diffraction to electron diffraction. Why must quantum theorists still ignore the quantum theory?

Summary

It certainly would be futile to discuss:

(1) the energy distribution over the spectrum of hermal radiation without taking note of Planck's rule for the energy exchange, $\Delta E = h/T$, of systems oscillating with *time period T*;

(2) the mechanics of atoms without taking note of Sommerfeld-Wilson's rule for angular momentum exchange, $\Delta p_\phi = h/2\pi$, for systems of *angular periodicity* 2π. But then it is likewise futile to discuss

(3) the mechanics of space-periodic bodies (crystals) without taking notice of Duane's quantum rule for linear impulse changes, $\Delta p = h/L$, in bodies with *space periodicity L*.

Whereas the first two quantum rules are recognized as the backbone of atomic theory, the third rule is generally overlooked. And yet it is the key to a simple and unitary explanation—in terms of pure particle mechanics and without appeal to a mysterious wave manifestation—of those diffraction phenomena which are usually quoted as supporting a dual aspect of matter. When Niels Bohr declared: 'The interference effects which appear when electrons pass through crystals are incompatible with the mechanical idea of particle motion', this cannot be accepted unless one blinds himself against the third quantum rule of mechanics. Nor is it true that the quantum rules represent links between the particle and the wave picture. The quantum rules (1), (2), (3) above are concerned only with particle mechanics, and not with joining two subjective pictures. Since Duane and Born, wavelike appearances of matter are to be interpreted as produced statistically by the same particles which have long been recognized as the substratum of all material phenomena. If particle mechanics is dominated by certain restrictive rules, the said quantum rules can be deduced from simple non-quantal postulates, as will be seen in later chapters of this book. The present Introduction is bent on showing that the phantom of waves versus particle appearances becomes obsolete under the powerful third quantum rule of mechanics. The way is thereby opened for a return from subjective pictures to the objective realism which has guided the progress of science for three centuries. I suggest that the Copenhagen School starts from the *wrong physics* when it maintains that wavelike appearances, in particular those of matter diffraction, are due to a periodic wave action of the electrons. Actually they are due to the periodic structure of bodies in space (crystals) and time (oscillators) *via* the corresponding quantum rules for the momentum and energy activity of the periodic bodies.

CHAPTER II

CAUSALITY AND CHANCE

'There is no cause and effect in nature; nature has but an individual
existence; nature simply *is*. Recurrences of like cases exist only in the
abstraction which we perform for the purpose of mentally reproducing
the facts.'

ERNST MACH

Quantum mechanics is a theory of probabilistic connections
between atomic data, obtained from macrophysical instruments and projected with the help of theory into the microphysical domain. It may be in order therefore to give a short
survey of the topic 'Causality and Chance in Modern Physics'.

8. *Laplace's determinism and Kant's causality*

The Laplacian doctrine of *universal mechanical determinism*
maintains that the world as a whole runs like a clockwork, its
present state being connected with every past and future state
by an unambiguous fixed relation by virtue of the mechanical
equations of motion. An all-embracing spirit viewing the
whole show would be able, after a short glance now, to predict and postdict the history of the world from the infinite past
to the infinite future. In the space-time picture of relativity
theory nothing ever 'happens', the world lines of all mass
points being fixed once and for all. However, Laplace's rigid
world actually disposes of causes and effects unless 'cause' is
taken as synonymous with 'before' and 'effect' with 'after':
post hoc and also *ante hoc, ergo propter hoc.* One could say
'Christmas is coming because the leaves are falling' as well as
'the leaves are falling because Christmas is coming'. Also,
conditions five billion years ago were such as to determine
every automobile fatality today. But this is an empirically
empty assertion since there is no room in Laplace's fixed
world to find out whether certain present variations would

23

always be followed by the same future variations. And most
of all, any present variation, if possible at all, would have to
change not only the future but the past as well, a rather dis-
turbing implication. One may argue: Since certain events
actually happen today, this shows that they *were* conditioned by
the state of the world five billion years ago. But this argument
only reveals that Laplace's determinism equates causal with
temporal sequences.

Laplace's determinism is related to *logical determinism*
which declares: What is, is; what has been, has been; and
what will be, will be. And nothing can change what will be to
what will not be. This again is not a testable law of nature; it is
a truism.

Another version is *fatalism* which maintains that what
will be in the future will be, although what is now can be
changed. All human activity now is futile. Neither medical
care nor any other human act can change anything: *Che sarà,
sarà*. Although fatalism claims that there is an iron determinism, it actually denies a link between what is done now
and what will happen later. It is full of inconsistencies.

Still we ask: What is the source of our persistent notion that
there is a *necessary* connection between events, denoted as
causality? To this question Immanuael Kant gave the following answer: The idea of a necessary or 'causal' connection is
the form, the framework in which we rationalize our sense
impressions to yield experience; it is not a part of experience
itself. We hold this idea before all experience, *a priori*. And
by virtue of its aprioristic character, *the causality idea is
immune to experience*, as reflected by the phrase: 'I shall never
bring myself to believe that anything happens without sufficient cause; if there is no demonstrable cause, then there
must be a hidden one.' This view, which no experience can
verify or falsify, is harbored by everybody including gamblers
and even quantum theorists off duty.

When physicists claim that 'modern science has repudiated
Kant', they refer to certain experiences in the atomic domain.

This view is held by Heisenberg([1]) who gives the following 'two answers to Kant': 'The one is: we have been convinced by experiment that the laws of quantum mechanics are correct. And if they are, we know that a foregoing event as cause cannot be found.' But Kant's idea or discovery is that causality is a judgment *a priori* inherent in our mind structure, a psychological fact so to speak. As such it does not depend on whether specific causes can be *found*. Every 'experiment' already rests on assuming causality. The special degree of causality, what the physicist calls 'individual' or 'statistical' according to current theories, has nothing to do with Kant. Heisenberg's second answer to Kant reads: 'If we wanted to know why an α-particle was emitted at that particular time, we would have to know the microscopic structure of the whole world including ourselves, and that is impossible. *Therefore*, Kant's arguments for the *a priori* character of the law of causality no longer apply.' This is debatable, too, because in the first place, no experiment or experience can defeat Kant's aprioristic argument. Second, physics is strained too: In order to know the instant of an α-emission in advance, if *any* knowledge would help, knowledge of a space 10^{-30} cubic-centimeter would do. Why the whole world? And the 'including ourselves' in the whole world does not make clear whether it means our physical bodies, which would be trivial, or whether it alludes to the fallacy (Chapter VIII) that atomic events, in contrast to macrophysical ones, become 'actual' only by entering the consciousness of a human observer. It seems that Heisenberg's 'two answers to Kant' in no way repudiate the principal tenet of the *Critique of Pure Reason* of the great philosopher Kant since, as R. D. Bradley([2]) puts it: 'The absence of an experience of causality is not equivalent to an experience of the absence of causality.'

The law of causality that nothing happens without sufficient and necessary cause, and more specifically 'when the cause (the present) is changed then the effect (the future) is also changed' becomes a testable rule which might either be right or wrong when Laplace's world as a whole 'including ourselves' is replaced by a section of the world as the object, and an outside agent who can modify the external conditions for the object at his will, study the ensuing changes, and regard them as 'effects' of his own doing. He must perform such experiments repeatedly in order to ascertain *regular* sequences. In order to eliminate the subjective element, the observer may interpose between himself and the object A a second object B as an instrument which interacts with A and signifies changes in A in an indirect, preferably magnified manner. However, in this case, the instrument B and the object A together merely constitute a larger object $(A + B)$ influenced by the observer. Within $(A + B)$ there is cause-effect symmetry: When a hot iron is put under a kettle, the iron may be said to cause the water to become warmer. But with the same right the water may be said to cause the iron to become colder. Only within an *asymmetric* situation, when the instrument B is taken as an extension of the observer's arm probing the object A, may this deliberate act be said to be the 'cause' of the change in the object A. In the words of Max Black:[3]: 'Cause acquires a meaning only as a conscious and voluntary act of a subject making something to happen to an object. . . . And anything having the tendency to show that the agent was not acting freely but responding to restraint . . . would immediately show . . . that *he* was not the cause but merely the instrument or an intermediary between the true cause and its effects.'

According to this version, then, causality is meant, not as a form of aprioristic rationalization as Kant had it, but rather as being connected with our notion of *free will*, of being free agents acting on the unfree rest of the world—a somewhat paradoxical psycho-physical theory. Unfortunately it leaves two contradictory answers to the eternal question: 'Can I will to will what I will?,' to which I myself as a subject reply: 'Yes, I can,' whereas someone else, in particular an educator or a public opinion manipulator will answer: 'No, you can't.'

9. *Causality as predictability*

For the scientist the issue of causality condenses to the question whether one and the same advance preparation of a finite system (rather than of 'the whole world including ourselves') is always followed by the same future development, so that on the grounds of many recurrences one may predict future events from present situations. Whether this is always possible is a matter of experience. The answer must not depend on aprioristic notions or theory. Of course, classical *theory* is deterministic, and quantum *theory* is not. But the experience in games of chance testifies to unpredictability, and has done so long before the quantum age.

Take the example of balls dropped through a chute aimed at the center of a blade. If, or since, a ball cannot indefinitely balance on the edge, it will have to make an asymmetric decision. Some balls drop to the right and others to the left in spite of equal advance preparation. A practically symmetric arrangement yields individually asymmetric results. Nevertheless, a Kantian, to whom causality is an aprioristic notion, will never renounce the idea of sufficient causes, if not a visible asymmetry, then a hidden one.

However, in spite of the unpredictability of future events in statistical situations, one often can *reconstruct* theoretical causes from the visible effects. When a ball ends up to the right in spite of central aim, one may ascribe it to an invisible air current which is unpredictable, too. And the deflection of electrons from a crystal, though individually unpredictable, can and will be ascribed to corresponding deflecting impulses during the collision.

Einstein's much quoted words: 'I cannot believe that the Lord plays dice with the world' meant that 'chance' only expresses temporary ignorance, whereas in nature everything has its place, is determined. However, the belief in determinacy is as much beyond the domain of physics as the opposite belief: that nature as such is indeterministic (as many quantum theorists maintain). Both are metaphysical assertions. Observation only shows that equal preparation, as far as we can achieve equality, leads to different unpredictable results, in Monte Carlo as well as in the atomic laboratory. This general experience can be elevated to a physical 'principle of uncertainty'. It is Heisenberg's great merit that he has established quantitative limits to the uncertainty of prediction. But uncertainty as such was known from every game of chance; it also was the background of statistical mechanics long before the quantum age. In contrast to prequantal experience, classical mechanical *theory* was deterministic and left no room for chance. In the quantum domain, however, experience and theory agree. For a clear discussion of the issue refer to R. D. Bradley([2]) and M. Bunge([4]), to name but two.

In view of the empirical unpredictability of individual results in games, ordinary as well as atomic, it is all the more remarkable that there are recurrent, hence *predictable averages* for unpredictable individual events. That is, a sort of causality prevails for statistical results. In both the ball-blade game as in the splitting of a ray of silver atoms into a north and south branch by a magnetic field (Stern-Gerlach effect) there is a 50:50 ratio corresponding to the symmetry of the arrangement with respect to the two sides. This *conformity between geometrical and statistical symmetry* is, on the one hand, an observed fact; on the other hand, it could be expected even without previous experience, as almost self-evident. Indeed, we would be very much intrigued if conformity should not prevail, and we would immediately blame some geometrical asymmetry for an observed statistical asymmetry, if not a visible one then a hidden one. The question thus arises whether the law of conformity of geometrical and statistical symmetry is indeed empirical, or aprioristic, or whether it is perhaps a way of defining the one 'symmetry' in terms of the other. Let us consider a concrete example.

Suppose the *r:l* ratio in the ball-knife game persistently comes out to be 49:51. As remarked before, our reaction will be that the aiming device *must* be slightly unsymmetric, and might be corrected by the turn of a screw. Does this not show that conformity between the two symmetries is not a law of nature but rather a matter of *defining* one symmetry in terms of the other? I do not think so. Conformity of the two symmetries is still an experience, although resting on approximate evidence only. Besides, even in case of an (approximately) central aim, three throws lead necessarily to an asymmetric $r:l$ ratio, either 2:1 or 3:0 or 0:3 or 1:2. Vice versa, even when the ratio in a thousand tests is exactly 50:50, we cannot be sure whether this is not merely a lucky accident in spite of a slightly unsymmetric set-up. We thus are led to the conclusion that, as an *approximation*, the conformity law is empirical. At the same time its *idealization* has the character of self-evidence rendering it apt to *define* one (exact) symmetry in terms of the other.

There have been other laws of nature which were regarded as empirical and *a priori* self-evident at the same time, such as the 'law' that every moving body necessarily comes to rest when left alone. Later its very opposite, Newton's law of inertia, became regarded as both empirical and self-evident, if not as a definition of 'being left alone'.

10. *The marvel of statistical co-operation*

Adherents of the deterministic ideology may still claim that each r- and l-event *must* have its distinct asymmetric cause, be it a slight deviation of aim, or a small perturbation of the ball during its flight, and that each such perturbation in its turn is the effect of an earlier cause. According to this view, then, an individual r-event is but the terminal member of a deterministic chain $r, \ldots rrr$, originating in the remote past and predictable by Laplace's spirit. Similarly, the date of a mortality could have been foreseen by an ideal doctor. And each single automobile accident is not an accident but has its definite cause, overt or hidden.

The trouble is that individual causality, instead of solving the paradox of co-operation of geometrical and statistical symmetry, rather aggravates it. Causal chains shift responsibility for the present 50:50 distribution occurring under approximate geometrical symmetry from today to yesterday and further to a remote past when there was no geometrical symmetry at all, when the atoms of the present gambling device were spread all over the globe. The observed conformity of the two symmetries today thus leads, if we adhere to the idea of continuous causal chains, to non-conformity in the past— unless we ascribe to each atom in the past an incredible amount of foresight or pre-ordainment of the rôle it will have to play as a member of any future statistical distribution.

We are faced here with a situation which, from the causal point of view, offers an insoluble paradox, as expressed by Bridgman[5] as follows: 'How can individual events, admittedly independent of one another, combine into a regular aggregate unless there is a factor of control over their combination? But what kind of control can there be over independent events?' Indeed, the co-operation of independent events forming predictable statistical averages as well as predictable average fluctuations away from the average, both computable *a priori* from the geometrical symmetry of the arrangement, is no less than a marvel. Remember that, in spite of the symmetric set-up, each single ball or atom has to make an unsymmetrical decision, going either to the right or to the left. Thus, what kind of control can there be to turn small number asymmetry into large number symmetry? There is no deterministic answer to these questions. Absence of plausible reasons for non-conformity of the two symmetries is not a causal explanation for their actual conformity. And plausibility is a tenuous argument, as testified by the stubborn belief of many laymen that, after seven heads in a row, it is propitious to bet on tails with more than even odds.

In view of this impasse, defenders of determinism have taken refuge either in hard or in soft determinism. All determinists agree with Einstein when he said 'I cannot believe that the Lord plays dice with the world'. If Einstein had added: 'nor that He has ever played dice', it would have been a clear and consistent statement of *hard determinism*. But how would a hard determinist account for the strange (from the deterministic viewpoint) fact that in all cases of statistical distribution, of gas molecules, of clicks in Geiger counters, of raindrops falling per square yard, etc., one always finds agreement between statistical observation and the mathematical theory of random, as to average distribution as well as to fluctuation? Does this not indicate that at least *once* there must have occurred a 'mixing' of conditions in which random was supreme? This is what soft determinists concede with their hypothesis of molecular disorder, supposedly created only once long ago and from thereon passed along deterministically through the ages. But if they admit a 'true mixing' of initial conditions at one remote instant, then there is no world-shaking step from there to the concession of *indeterminism* that such mixing may take place at other times, too, whenever a new statistical situation in agreement with the mathematical theory of random occurs, in gambling casinos, in 'classical' molecular distribution, in quantum processes such as radioactive disintegration and so forth. It also may be remembered that statistical co-operation of independent events, often regarded as a matter of course, originally was an important empirical discovery by our remote ancestors. It marked a significant step forward in natural philosophy from the creed in a whimsical and hostile world, toward a *cosmos* in which even the god of thieves and gamblers is bound to obey rules of fair play. And what is fair play in a game? Conformity between geometrical and statistical symmetry! Fair dice have equal faces which claim equal chances. Yet, although conformity of geometry and statistical patterns is plausible, there is no causal explanation for it. A precarious way out of these difficulties may be seen in the following hypothesis of soft determinism:

Once upon a time there was a *demon* who, in the ball-knife

game, first started two causal r-chains, then one l-chain, then three r-chains. Thereupon, realizing that he had given too much preference to r-chains, he started four l-chains in a row, deliberately causing a pseudo-random distribution in order to lure present-day physicists away from the true deterministic faith and deceive them into accepting the heresy that the Lord plays dice with the world, or at least has played dice once long ago. In contrast to Maxwell's good demon who turns molecular disorder into order, the demon described here is evil because he deliberately creates disorder. To help his evil work, he may use the even and odd sequence of decimals of the number π (which in fact are orderly so as to present π, and only look randomlike), or the method proposed by K. Popper([6]) for producing a randomlike series. However, the hypothesis of a *deus ex machina* as an excuse for apparent disorder is not a very good argument to save determinism.

11. *Similarity of ordinary and quantum games*

As seen before, the hypothesis of unobservable concealed causes for statistical co-operation does not leave room for a consistent deterministic ideology. This applies to all cases of individual disorder producing statistical order, for dice and ball-blade games as well as for the two-way deflection of atoms and radioactive disintegration. I therefore can see no justification for the claim of many quantum theorists that only their branch of science, and the Heisenberg uncertainty rule in particular, has brought defeat to the doctrine of uninterrupted causal chains. If they have been converted to indeterminacy only by Heisenberg's rule, and have not seen the inconsistency of combining the hypothesis of casual chains with the actual existence of random distribution, this is their fault. To me it seems that *every* statistical ensemble proves the failure of individual determinacy and shows that we may regard acausality (or unpredictability, using a subjective term) as an autonomous principle of physics. There is no factual difference in this respect, no matter what one thinks about it theoretically, between ordinary and atomic games of chance. Consider the two following parallel experiments: (1) A great many vessels, all of the same size and each containing one small ball or molecule as well as an escape hole, and (2) a great many radium atoms, each containing an α-particle. In both cases one observes an exponentially decreasing average number of escaping particles per second. An unbiased observer will be hard to convince that the α-escape, dominated by the statistical laws of quantum mechanics, should be of great philosophical significance in showing an irreducible acausality, whereas the escape of particles from a vessel does not. Two generations ago a difference might have been seen in the fact that the α-escape rate seemed immune to external influences, whereas the escape rate of molecules can be controlled by varying the temperature and the size of the hole. This difference has evaporated since one has learned to smash atoms in a statistically controlled manner. Therefore I cannot see why the uncertainty of the instant of an atomic disintegration should be *essentially* different from the uncertainty of the escape instant of a gas molecule from a vessel. Bohr attributes the gas escape *only* to the complexity of the situation and to our ignorance of the initial conditions, whereas he regards the radioactive escape as 'essentially' acausal. There is a difference indeed between experimental facts and what one thinks about the facts. About many molecules in one vessel or about many vessels each containing one molecule, Bohr thinks deterministically; about α-escape he thinks acausally. The same empirically unsupportable distinction is made when it is said that 'in ordinary games of chance we *do* not know, whereas in atomic quantum games we *can* not know'. It is a distinction without a difference and illusory on empirical grounds, because deterministic theory is always inconsistent with statistical order, demons being excluded.

Although prediction is not possible in games of chance, one may *reconstruct* individual causes. But again, there is no basic difference between ordinary and atomic situations. When a ball falls to the left of the blade, one may attribute it to a corresponding whiff of wind. When an electron diffracted through a crystal is deflected into an n'th order maximum, one may ascribe it to a corresponding impulse, $\Delta p = nh/L$, given off by the crystal to furnish the required deflection. Those individual hypothetical causes are *ad hoc* invented for the purpose of reconstruction. But special causes (why just nh/L with a special n?) lead to an infinite regress in both cases, ordinary as well as atomic. For this reason I cannot follow Niels Bohr([7]) when, in his famous 'Discussion with Einstein', he writes: 'It is most important to realize that the recourse to probability laws under such [atomic] circumstances is essentially different in aim from the familiar application of statistical considerations as practical means for accounting for the properties of mechanical systems of great complexity. In fact, in quantum theory we are presented not with intricacies of this kind, but with the inability of the classical frame of concepts to comprise the peculiar feature of undivisibility or individuality, characterizing the elementary processes.' In my opinion, it is in the contrary most important to realize that there is no essential difference between statistical phenomena in ordinary games and in case of gases consisting of many particles, as against atomic experiments, between the escape of balls from a large box with a small hole as against the escape of α-particles from a radium atom with a 'hole'. Those who have thought critically about statistical mechanics have inevitably found that one *always* has to introduce an element of disorder, a basic uncertainty for individual events, into the theory. The innovation of quantum theory is to have established quantitative limits for the uncertainty of special physical quantities p and q (Heisenberg). But it is misleading to say that in ordinary games and in gases 'we could if we would' restore determinism, whereas in quantum games 'we could not even if we would', or also that in case of ordinary games there are 'accidents for us' and in atomic games there are 'accidents *an sich*'. And when P. Bridgman([5])

remarks that 'the seeds and sources of the ineptness of our thinking in the microscopic range are already contained in our present thinking in the large-scale region and should have been capable of discovery by sufficiently acute analysis of ordinary commonsense thinking', I agree: The idea of getting along with determinism in ordinary but not in atomic games shows an ineptness of thinking indeed. Classical determinism is helpless in the face of statistical contingencies. The insight that hypothetical concealed causes lead nowhere has been gained already in pre-quantum days, in opposition to the claim of Bohr in the sentence quoted above.

12. *Molecular disorder*

Deterministic theory can never account for statistical law without admitting the non-deterministic hypothesis of 'molecular disorder' occurring in every new game of chance. This has been realized already by the founders of classical statistical mechanics. Yet it is sometimes forgotten, for example when even Einstein maintained, in an exchange of letters with K. Popper([6]) that 'it is possible to derive statistical conclusions from a deterministic theory', his example being a mass point running with constant velocity along the periphery of a circle, divided into a right and left half. It is true that the *r:l time average* is 50:50 according to deterministic theory. Nevertheless Popper is right that there are no deterministic reasons why one should find in many inspections a 50:50 *statistical average* ratio unless one relies on the additional hypothesis of 'molecular disorder', applied here to the instants of inspection. The statistical result is not derived from deterministic theory but from the assumption that the inspection instants display statistical symmetry. The same holds in more complex situations. For gases consisting of many atoms, the ergodic theorem allows us to calculate space averages at one time and time averages at one place. But there is no deterministic reason why averages over continuous space and time should agree with averages taken at random instants and at random places. Appealing to the hypothesis of molecular disorder is no more than begging the question.

Since these considerations apply to dice games and gases as well as to atomic quantum games, it is hard to see much sense in an extra proof that the quantum events in atomic dimensions have an irreducible character of uncertainty, or that the formalism of the quantum *theory* requires a non-deterministic interpretation. A given formalism can always be interpreted in a variety of ways. Von Neumann's[8] proof to the contrary is circular, as shown by de Broglie[9] and Feyerabend[10], hence it is useless. But it is likewise delusive to think that the refutation of von Neumann's proof has reopened the path of determinism into quantum physics. No statistical situation can ever be reduced to determinism, unless one believes in a *deus ex machina* who at one remote time has introduced statistical disorder which then is passed on deterministically to the present and future. We here agree with Charles C. Peirce[11] (*The Doctrine of Necessity Examined*, of 1892, long before the quantum age) in reply to a soft determinist: 'You think that all the arbitrary specifications of the universe were introduced in one dose in the beginning, if there was a beginning. But I for my part think that the diversification, the specification has been continually taking place.' For a recent evaluation of the internal contradictions of the deterministic doctrine refer to K. R. Popper, 'Indeterminism in Quantum Physics and in Classical Physics'.[12] See also *Determinism and Freedom in the Age of Modern Science, a Philosophical Symposium*[13].

Summary

In contrast to the purely academic Laplacian doctrine of universal determinism, and in contrast to Kant's causality as an aprioristic form of rationalization, physicists speak of causality when a state A of an object, frequently prepared in the same manner, is always followed by the same state B so that one can *predict* B to follow A also for future experiments.

Whether this holds in all cases is a matter of experience. Every game of chance refutes it; *A* is sometimes followed by *B* and on other occasions by *B'* or *B"*.

Different effects *B*, *B'*, *B"* following *identical* preparation *A* (as far as practically possible) occur in (practically) fixed statistical ratios. In the special case of geometrical symmetry of the gaming device, statistical symmetry is observed, too. This *conformity* of geometrical and statistical symmetry is (a) an empirical fact, hence of approximate confirmability only, (b) expected as an exact idealization before all experience, (c) a criterium to define geometrical symmetry in terms of statistical symmetry, or vice versa.

The general observation of *statistical co-operation* of individual events in forming statistical patterns leads to a breakdown of the deterministic idea in principle. Individual concealed causes only shift the blame for today's statistical co-operation to yesterday, and then to an infinite regress. Individual causality could at best be saved by the 'quagmire of soft determinism' according to which one may assume a *demon* who at some remote time has deliberately produced a pseudo-random distribution. Essentially the same is assumed by the term 'molecular disorder' which merely begs the question. Unpredictability, denoted as acausality of individual events, must be accepted as an irreducible feature of natural science. Quantum theory has not contributed to this insight in general; but it has drawn definite quantitative limits of uncertainty for atomic games of chance. Unpredictability of future events does not preclude the *reconstruction* of individual causes on the grounds of deterministic theory.

CHAPTER III

CONTINUITY AND QUANTUM JUMPS

'When the cases approach each other continuously and finally get lost in one another, then the events in the sequel do so also.'

LEIBNIZ

13. *Statistical law and continuity principle*

In the preceding chapter we saw that the fundamental fact of independent events co-operating to form statistical patterns, whether macroscopic or atomic, can *never* be understood on a deterministic basis. The claim that only quantum experience has discredited determinism is unwarranted therefore. It may be of interest, however, that there is another, certainly prequantal, principle from which the occurrence of statistical distribution can be derived as a necessary consequence. We refer to the principle of *cause-effect continuity*, first enunciated by Leibniz, in the words chosen as the motto of this chapter, or in modern parlance: 'An infinitesimal change of cause never produces a final change of effect.' We here quote Leibniz's own example which served him to correct an impossible speculation by Descartes concerning the kinematics of elastic collision. According to Descartes, when two balls collide with equal but opposite velocities, both balls will, after the collision, pursue their path in the direction of the heavier ball. But if both balls have the same weight, they will rebound in opposite directions. According to Leibniz this cannot be so since it violates the continuity principle: there must be a *gradual* change of the final velocities and directions when the weights of the two balls are gradually made more and more equal; in particular when the weights are *almost* equal, the final velocities must be *almost* equal and opposite, rather than both in the direction of the heavier ball.

38

Let us next apply the continuity principle to our ball-blade game. When the aim is at a considerable angle α to the right, all balls fall to the right of the edge. The same holds for the left. Instead of a game of chance we have certainty. But imagine that the physical aim is changed from right to left in a gradual manner. From an idealized point of view one may expect that, at the moment when the angle α crosses zero, an abrupt change from all *r*-balls to all *l*-balls, from a 100:0 to a 0:100 ratio of *r:l* will take place. The general principle of continuity tells us, however, that this cannot be so, that we rather ought to expect a gradual change, from the 100:0 to the 0:100 ratio, when the physical aim is gradually shifted through a small but *finite* range $\Delta\alpha$ enclosing $\alpha = 0$. What else can those intermediate ratios be than *statistical* ratios, since every single ball can only fall either to the right or to the left of the blade! Within the range $\Delta\alpha$ there can be no certainty of the outcome; it must be a game of chance. We see here that *statistical distribution*, with intermediate ratios between 100:0 and 0:100 may be regarded as a *rational consequence of the principle of cause-effect continuity*. Again there is no difference between ordinary and atomic games. Here and there, continuity requires that the same set-up as cause, if it leads to different events as effects, will produce statistically ordered ratios. Refer to V. Hinshaw's critique[3].

There *seem* to be contradictions to continuity in the thermal behavior of substances which, upon infinitesimal change of temperature, are supposed to change their state of aggregation abruptly. However, although this may be so according to idealized theory, actually the change takes place within a final range ΔT of temperature T. And it is significant that within this range statistically controlled fluctuations take place.

The close connection between overall cause-effect continuity on the one hand, and individually unpredictable 'acausal' though statistically ordered events on the other, ought to be of interest not only to the physicist but to the philosopher of science as well. As just seen, the continuity

principle offers *sufficient reason for the lack of sufficient causation.*
Moreover, when Leibniz wrote([1]): 'The principle of continuity is beyond doubt to me and might help to establish
several important truths in a genuine philosophy [= science]
which rises above the imagination to seek the origin of
phenomena in intellectual regions', his prophetic words are
vindicated by the possibility of constructing modern quantum
mechanics on the basis of his principle, as the following sections may show.

14. *Separation of atomic states*

Suppose many atoms of the same species are all in the same
state A, that is, they have all reacted with Yes to a certain
selective instrument which defines the state A. Let us call this
instrument an A-passing filter, or an A-filter for short.

When A-state atoms are sent once more toward an A-filter,
they will pass again (Fig. 2a). However, the same atoms may
also possess other states $\bar{A}$ which are repelled by the A-filter.
We denote them as states entirely different from A, written $\bar{A} \neq A$, and separable from the A-states by means of
the A-passing (and $\bar{A}$-rejecting) filter (Fig. 2b). For example,

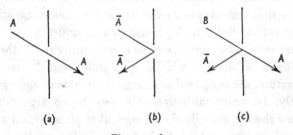

(a)   (b)   (c)

Fig. 2 a, b, c

when A denotes the spin orientation state $\uparrow$ of silver atoms,
the opposite orientation $\downarrow$ represents a state $\bar{A} \neq A$. Similarly,
the two orientation states $\nearrow$ and $\swarrow$ are separable by means of
a Stern-Gerlach field of direction $\nearrow$, characterizing the two
states $\nearrow$ and $\swarrow$ of silver atoms as entirely different, or separ-

able, or 'mutually orthogonal' in the quantum terminology. Any two energy states are entirely different, are 'mutually orthogonal', are separable by means of an 'energy filter' of some sort (all this is idealized, of course).†

It could have been expected from the continuity principle, and it is indeed confirmed by experience, that there is a third class of states intermediate between A and $\bar{A}$, between always passed and never passed, namely states B which are sometimes passed and at other times are repelled by the A-passing filter (Fig. 2c). The states B then can neither be regarded as equal to A, nor as entirely unequal, that is as 100 per cent separable from A. We denote such states B as 'fractionally equal' to A, written $B \sim A$. And we define the *fractional degree of equality* between A and B as the statistical *passing fraction* of B-state atoms through an A-filter. However, since the equality concept is mutual (which twin is more equal to the other?), this definition of fractional equality makes sense only when B-state atoms pass through an A-filter with the same probability as A-state atoms pass through a B-filter. This *symmetry postulate*, written

$$P(B \to A) = P(A \to B),$$

is quite fundamental for the probability metric to be developed later (Chapter VI). It is a quite plausible postulate since it is the statistical counterpart to the *reversibility* of processes in classical mechanics. The P-symmetry is decisive for quantum mechanics.

The phenomena of Fig 2c are known to the physicist as the *splitting effect* in a general sense; they are illustrated by the

† To the remark that a 'filter' is here defined with reference to 'states', and 'states' with reference to 'filters', I have this to say: A theoretical structure works with mathematical and verbal symbols without specific content. Thereafter it may be associated with a 'model', a physical counterpart. The classification of 'states' developed here is exemplified in quantum physics by 'states of position at a given time', by 'states of energy', and so forth, but certainly not by states of health or states of knowledge of some observer. The formal structure may be built without always referring to familiar experiments—just as the geometer applies the concept of points and straight lines without always giving an operational definition of these terms.

Stern-Gerlach division of an incident ray of magnetic particles in two deflected rays by means of a magnetic field. The term 'filter' is meant in a quite general sense. It refers to any instrument devised for distinguishing, that is separating, a state A from another state $\bar{A}$ entirely different from A. For example, a yardstick with a mark at a place with coordinate q together with a clock showing the time t is a (q, t)-filter in so far as it indicates whether a particle travelling along the yardstick is, or is not, at the place q at time t. If the yardstick is graded with a set of marks q_1 q_2 ... then it is a *separator* which distinguishes between various mutually exclusive (termed 'mutually orthogonal') states of position at a given time t. A Nicol prism is a *filter* for one definite state of polarization. A doubly refracting crystal is a *separator* for two mutually exclusive states of polarization of 'photons' (if one believes in photons).

Returning to the question of causal chains: If we had to do only with events belonging to *one* statistical ensemble, the hypothesis of concealed causes for individual events would be acceptable. The reason that most theorists, myself included, prefer to speak of *acausality* is the frequent occurrence of several statistical contingencies in succession. Suppose the passed B-state atoms had certain concealed causes for passing the A-filter, and the repelled ones for being repelled. That is, suppose the incident atoms actually are distributed over two different states, B_A and $B_{\bar{A}}$, predestined to pass or to be rejected respectively by an A-filter. If this were so then the A-filter would be the most convenient means of bringing the concealed difference between B_A and $B_{\bar{A}}$ into the open. Suppose now the passed B_A particles are tested by a C-filter; the latter will divide them again into a passed and a blocked fraction. According to the hidden-cause hypothesis, this would reveal that not even the state B_A is uniform but consists of a state B_{AC} and of a state $B_{A\bar{C}}$ at a definite statistical ratio, the former being passed and the latter rejected by a C-filter. Continuing in this fashion it becomes obvious that the concealed-cause hypothesis would ask us to believe *first* that every single atom is predestined for a definite reaction in any possible future filter test, and *second* that the fates of various atoms are co-ordinated in advance so as to yield definite statistical ratios (plus Gaussian fluctuations) of passed and blocked numbers, all this in harmony with mathematical error theory. It would require a foresight of really fantastic scope, a gigantic fraud staged by an evil demon bent on feigning random when there is concealed order. *One does not need von Neumann's proof* to see that in the presence of any statistical distribution whatsoever, quantal or pre-quantal, causal chains lead no-where. This should dash all hopes that individual causality might be restored on a deeper sub-quantum level.

15. *Reproducibility and quantum jumps*

When among 100 incident B-state particles 70 are passed by an A-filter and 30 are rejected, it is futile to look for individual differences in the B-states (perhaps to be denoted as B_A and $B_{\bar{A}}$), that is for the *why* of the different behavior. We may ask, however, *how* do the 70 particles manage to pass a filter built for passing A-state particles at a 100 per cent rate and to reject none? The answer can be found from testing the 70 passed particles by another A-filter: they now pass the A-filter without exception. From this experience—'once passed, always passed' and similarly 'once rejected, always rejected'—can be drawn the conclusion that those 70 incident B-state particles which are passed by the first A-filter must, in the act of passing, have turned or 'jumped' from their original state B to the new state A; similarly, those rejected by the first, and then also by the second A-filter, must have jumped from the state B to $\bar{A}$. This is indicated on the passed and reflected side of Fig. 2c. Similarly, using the particle theory of light: a photon which has once passed a Nicol prism will pass another parallel Nicol prism with certainty; hence it must have acquired the state of polarization parallel to the two Nicol prisms in the act of passing the first. The experience 'once passed, always passed' and 'once rejected, always rejected', the *reproducibility of a test result*, leads immediately to a typical quantum feature, a *jump* from the original state B to the new state A or $\bar{A}$ during a test with an A-filter, with *transition probabilities* $P(B \to A)$ and $P(B \to \bar{A})$ respectively. Such unpredictable jumps may thus be seen as a consequence of the elementary general experience of reproducibility, without relying on any special quantum assumption. Vice versa, A and B and C qualify as 'states' within the theory only when they satisfy the P-symmetry and the reproducibility requirement.

Classical *statistical* mechanics is based on the tacit assumption that once upon a time a random distribution had been established (by a demon?), and that only from there on all events were determined. Quantum theory maintains that every test involves a new independent statistical distribution. The difference of the two standpoints may be illustrated by the following example.

Imagine a great number, N, of arrows distributed as radii of a circle in a random fashion. Suppose this 'star' of arrows is viewed through a circular glass A divided into three equal sectors A_1 A_2 A_3. Accordingly, the betting odds of finding an individual arrow within section A_1 or A_2 or A_3 amount to 1/3 each. The statistical predictions in this and similar games are based on the initial random distribution of the arrows over all directions; and these predictions may be confirmed simply by *viewing* the arrows, without thereby affecting the state of any arrow so viewed. This is the classical situation.

When the N macroscopic arrows are replaced by micro-arrows representing the spin directions of Ag-atoms, then not even a demon can distribute them over all directions of a circle simultaneously. Spin arrows can be laid out only by passing them through a linear 'filter', for example a magnetic field held in a certain direction, from which the spins emerge in two opposite directions, parallel and anti-parallel to the field, according to chance. When the magnetic field is held in

the N-S direction, the *Ag-atom*s line up their spins in the
N and S direction only. When they are later tested by a
second field, this time of NE–SW direction, each spin is
compelled to *jump* from its original N or S direction either to
the NE or to the SW direction. For an individual spin it is a
game of chance whether it will jump to the NE or to the SW
direction; only the average frequencies of the various jumps
have definite magnitudes. (They are equal to the cosine
squares of half the angle between the original and the final
direction.) The macroscopic testing instrument, in the present
instance the magnetic field, plays an active part in compelling
the tested particles to jump from their original direction to
one of the two directions peculiar to the magnetic field. Which
of the two directions is reached by an individual particle is
unpredictable in principle. Similarly, when a position meter
has located a particle at a certain point with coordinate q,
or within a small range δq, and when this observation is followed by a test of the momentum of the particle, then it is
unpredictable which momentum p will actually turn up. No
experimental arrangement can put a particle into a state with
definite position and momentum simultaneously in a predictable fashion (Heisenberg). The smaller the accuracy of
location δq, the larger will be the uncertainty of the momentum p acquired in the act of locating. This does not prevent
us, however, from afterwards *reconstructing*, from the observed deflection of the particle, the momentum acquired by
it with an accuracy far surpassing that of the range $\delta p \sim h/\delta q$
of prediction. And since measurement always involves reconstruction of what has happened, it is not justified to maintain
that Heisenberg's rule concerns the impossibility of simultaneous exact measurement of a qp-pair. Merely, the result
of a measurement cannot be prepared and predicted. More
about this in Section 37.

It has been argued, first by Reichenbach([2]), that we are
faced here with a new logical situation: Besides the contrast
between Yes (passed) and No (repelled) there is a third case

Undetermined. Reichenbach has even constructed a schema of a three-valued logic. To me this seems establishing another contrast of incontrastibles: The Yes and No refer to facts, whereas the Undetermined refers to what one may *think* about future facts which will turn out to be Yes or No. *Causal anomalies are not logical anomalies.* Rather, the occurrence of acausal events controlled by statistical law is foreshadowed by the postulate of cause-effect continuity. And the apparent paradox that the *continuity* postulate together with that of reproducibility requires *discontinuous* jumps from state to state, from B to A or $\bar{A}$ respectively (Fig. 2c), is removed by the realization that those discontinuous transitions from state to state in tests are controlled by chance rather than by cause-effect relations. We thus have macroscopic cause-effect continuity entailing microscopic acausal discontinuity.

In order to understand quantum theory it is appropriate to begin with an analysis of such simple experiments as the splitting effect, rather than starting from Planck's $E = h\nu$ which is a remote consequence of several elementary postulates in combination. A simple straightforward approach to the intricacies of quantum mechanics has been obstructed, however, by the historical development which derived its impetus from spectacular discrepancies between microphysical fact and classical expectation, rather than from a critical analysis of the ineptitude of classical deterministic thinking about 'ordinary' games of chance. With the sole exception of Bohr's correspondence principle, the emphasis has always been on the dramatic and revolutionary; and it had to be so in order to justify the break with the Cartesian tradition. Then at long last it all seemed to condense in the Schrödinger equation, today placed at the entrance gate to quantum theory. But this is like beginning a study of mechanics with $E = mc^2$ rather than with falling apples and swinging chandeliers, or electromagnetism with the Maxwell field equations rather than with Coulomb's law. Intricate developments involving complex imaginary ψ-functions periodic in coordinates and momenta according to $p = h/\lambda$ and $E = h\nu$, ought to be placed at the end, not at the beginning of the theoretical exposition. I therefore have reversed the usual line of approach. Our first concern is to develop the general schema of connection between probabilities. Special applications to conjugate observables q and p are left to Chapter VII, where the wavelike h-dominated features of quantum dynamics are developed.

Summary

The principle of cause-effect continuity, first enunciated by Leibniz, may be applied to situations which degenerate into 'games of chance' when the two reactions—always A or always $\bar{A}$ (meaning non-A)—are bridged by sometimes A and sometimes $\bar{A}$ at a statistical ratio between the limiting cases of 100 per cent (always A) and 0 per cent (always $\bar{A}$). A particular game of chance occurs in the testing of 'states' of a system by means of filters in the most general sense. When a B-state passing filter always rejects the state A, then A is entirely different from B, written $A \neq B$. When B always passes A, then $A = B$. But there must be, according to the continuity principle, intermediate cases of *fractional equality* between A and B, written $A \sim B$, where B-state particles are sometimes rejected, sometimes passed by an A-passing filter. The passing fraction $P(A \to B)$ of B-particles passing the A-filter may then be taken as the quantitative definition of the equality degree between states A and B. This definition makes sense only when $P(A \to B) = P(B \to A)$, and when 'states' are defined so that they fit into this schema of symmetric passing fractions.

When one introduces the postulate of reproducibility of a test result then the passing fractions become *transition probabilities* from A to B and *vice versa* in tests. The principle of cause-effect continuity for averages, symbolized by the relation $A \sim B$ of fractional equality, gives sufficient reason for the absence of sufficient causes for individual choices of events belonging to a statistical ensemble. Microphysical acausality, that is individual unpredictability, may be regarded as following from the principle of cause-effect continuity in the macroscopic domain, as examplified by the statistical effect of a continuous change of aim in the ball-blade game.

CHAPTER IV

TRANSITION PROBABILITIES

'The structure of the system is the work of reason; but the empirical contents and their mutual relations must find their representation in the conclusions of the theory.'

EINSTEIN

16. *Two-way symmetry*

The previous chapter dealt with the relation between only *one pair* of states, A and B, in the three cases of equality, inequality, and fractional equality. Similar considerations can now be applied to *all* states of a microphysical object[1].

A given atom may be tested for its position at a certain time (q, t), then for its energy E, its electric moment, for other observable quantities, generally denoted as A, B, C, and so forth, by bringing the atom under the scrutiny of an A-meter, B-meter, etc. The A-meter will reveal the atom to be in *one* among many possible states $A_1 A_2 A_3 \ldots$ For the sake of simplicity let us assume that our atom has only a finite number of possible A-values numbered $A_1 A_2 A_3 \ldots$ A_M. M is called the *multiplicity* of the A-states. Similarly, under a B-meter test our atom may display a finite number of values $B_1 B_2 \ldots B_N$. All this is quite trivial.

Far from trivial, however, is the experience that there are certain qualities, denoted as *observables* in the narrower sense of being connected with one another by definite *probability relations* in the following manner. Suppose the atom, under an A-meter test, has displayed a particular state A_k. When the atom is now subjected to a B-meter test, then any of the states $B_1 B_2 \ldots B_N$ may turn up, without the possibility of predicting the result of an individual B-test. A series of tests, however, always starting from the same initial state A_k (previously ascertained by an A-meter) may yield the various

B-states in a statistical distribution with definite relative frequencies or probabilities, $P(A_k \to B_j)$, the sum of which is unity:

$$\sum_j P(A_k \to B_j) = 1 \qquad \text{for } k = 1, 2, \ldots M. \quad (1)$$

Suppose now that the atom in the original state A_k, when subjected to a B-test, ends up in a certain state B_j, whereupon the atom is subjected to an A-test again. It turns out now that the atom has lost all 'memory' of its original state A_k. Instead, a series of A-meter tests applied to the state B_j yields various A-values in a definite statistical distribution, independent of previous history, with probabilities $P(B_j \to A_k)$ whose sum is in unity:

$$\sum_k P(B_j \to A_k) = 1 \qquad \text{for } j = 1, 2, \ldots N. \quad (1')$$

The transition probabilities P leading from initial A-states to final B-states may be compiled in a rectangular table or 'matrix' containing M rows and N columns:

$$\begin{pmatrix} P(A_1 \to B_1) & \ldots & P(A_1 \to B_N) \\ \ldots & & \ldots \\ P(A_M \to B_1) & \ldots & P(A_M \to B_N) \end{pmatrix} = (P_{\overrightarrow{AB}}). \quad (2)$$

Every row sums up to *unity* according to (1). Similar tables or matrices may be drawn up for other statistical series of tests yielding matrices $(P_{\overrightarrow{BA}})$ and $(P_{\overrightarrow{AC}})$ and $(P_{\overrightarrow{CA}})$ and so forth. Their rows always sum up to unity.

A special case is that of applying the same meter twice. When an A-meter test has shown an atom to be in the state A_k, and the A-test is applied to it again, the former result A_k will be found again with certainty, that is with probability *unity*, so we have

$$P(A_k \to A_{k'}) = \delta_{kk'} = \begin{array}{l} 1 \text{ for } k = k' \\ 0 \text{ for } k \neq k' \end{array} \quad (3)$$

A counter-example *seems* to be that of a particle, found at place q at time t, and found at another place q' at time t'.

This overlooks that position q is not an 'observable', but (q, t) is. A certain position can be found by a (q, t_1)-meter at time t_1. A (q, t_2)-meter, however, is not 'the same meter' in the quantum sense. Although $P(q, t_1 \to q', t_1) = \delta qq'$, there is a finite probability connecting (q, t_1) with (q', t_2), denoted as $P(q, t_1 \to q', t_2)$. Owing to the analogy of (q, t) with (E, p) in mechanics (E = energy, p = momentum), one would expect that E alone does not characterize a 'state' in the quantum sense, but that only (E, p) does. This is quite correct if one counts as p the kinetic momentum, mv together with the 'potential momentum'. Just as only the total = kinetic + potential energy is conserved, so is the 'total momentum'. (For details refer to books on mechanics.)

We come now to the theorem of a *two-way symmetry* of the transition probabilities:

$$P(A_k \to B_j) = P(B_j \to A_k). \tag{4}$$

In words: the probability of an atom in a state A_k, ascertained by an A-meter, jumping to a state B_j under the scrutiny of a B-meter equals the probability of the inverse process. This corresponds to classical deterministic mechanics where the development from state A to B is reversible, or symmetric with respect to the initial and final state.

It is important to realize that the symmetric transition probabilities P discussed here, pertaining to changes under test instruments (idealized) are in no way related to the transition probabilities which played a major part during the early years of the quantum theory. At that time one was interested in the probability $w(E \to E')$ of an atom jumping from a higher energy level E to a lower level E' by *emission* of a photon of energy $\epsilon = E - E'$ contained in radiation of frequency $\nu = \epsilon/h$, and of the inverse process of *absorption*. In case of thermal equilibrium between atoms and radiation, the two probabilities w must be equal. In order to obtain the Planck radiation formula for the spectral energy distribution in radiation, Einstein made the assumption that the absorption probability is proportional to the number N' of atoms on the lower level E' ready to absorb and to the number n of photons $h\nu$ in the radiation field *before* the absorption act, and that the emission probability is proportional to the number N of atoms in the higher level E as well as to the number $(n + 1)$ of photons *after* the emission act, so that he arrived at the equality

$$w(E \to E') = \text{const } N \cdot (n+1) = \text{const } N' \cdot n = w(E' \to E).$$

Since the atoms have the numerical equilibrium ratio

$$N/N' = \exp(-E/kT)/\exp(-E'/kT)$$

at absolute temperature T, k being the Boltzmann constant, and since

$$E - E' = \epsilon = h\nu,$$

the last three formulas yield the result

$$n = 1/[\exp(h\nu/kT) - 1]$$

for the average number n of photons of energy $h\nu$ occupying a radiation component of frequency ν, that is Planck's radiation energy law. The probabilities w refer to the transitions from one to another atomic energy level under the external *perturbing influence* of radiation, so that $w(E \to E')$ has a finite value. Our probabilities P above refer to transition under *tests* with (idealized) measuring instruments or 'meters'. P refers to one atom, w depends on the numbers N and n.

The two-way symmetry theorem, or postulate, or axiom (4), has far-reaching consequences. *First*, the columns of the matrix $(P_{\overrightarrow{AB}})$ are now identical with the rows of the matrix $(P_{\overrightarrow{BA}})$. But since the latter as rows sum up to *unity*, the same must hold for the former. That is, in supplement to the sum rule (1) we arrive at

$$\sum_k P(A_k \to B_j) = 1 \qquad \text{for } j = 1, 2, \ldots N. \quad (5)$$

That is, the sum of the probabilities of an atom arriving in one state B_j under B-meter tests from various initial states $A_1 A_2 \ldots A_N$ is *unity*. Because of the two-way symmetry, the arrows → can now be replaced by double arrows ↔, or simply be omitted from here on.

Second, if each row and each column of (P_{AB}) sums up to unity, it follows that there must be as many rows as columns. Indeed, suppose the set A has M members and the set B has N. When adding all P's columnwise, the sum is N; adding row by row, the sum is M. Hence $M = N$. Similarly, the rows and columns of the matrix (P_{BC}) must have the same multiplicity, and so forth. Hence all sets of states, $A, B, C \ldots$ of a given mechanical system must have the same multiplicity. In other words, the P-tables are squares with M rows and M columns, each of them summing up to unity. The P-tables may thus be denoted as unit magic squares or *unit sum matrices*. The technical name is 'doubly stochastic matrices'. The special case (3) indicates that (P_{AA}) is the matrix *unity*, with *ones* in the diagonal and *zeros* outside.

Equations (1) to (5) represent the formal background on which rest considerations concerning a mutual dependence of the various matrices (P_{AB}), (P_{BC}), etc., in particular the law of probability interference via probability amplitudes denoted as ψ (Chapter VI). The remainder of the present chapter is devoted to clarifications and applications.

17. *States and observables*

When atoms, originally in the state A_k, are tested with a B-meter they may turn to any one among the states $B_1 B_2 \ldots$ There are instruments called *separators* which distribute the incoming atoms directly over the B-states with probabilities or relative statistical frequencies $P(A_k, B_j)$. An example is the Stern-Gerlach separator which produces the splitting effect into various states of orientation in a magnetic field.

By blocking all components $B_1 B_2 \ldots$ save B_j, the *B-separator* can be turned into a B_j-*filter*. In case of the Stern-Gerlach B_j-filter consisting of a magnetic field of

B-direction, the passing B_j-state particles may be directed toward a Stern-Gerlach field of C-direction, enclosing a certain angle with the B-direction. In this case the B_j-particles, when passing the C-field, have to choose between the states $C_1\ C_2\ \ldots$ of orientation in the C-field; and they carry out the 'jumps' from state B_j to the various states C_n with probabilities $P(B_j, C_n)$. States of orientation B_j and C_n are called mutually *incompatible*. If a particle, originally belonging to the B_j-group, is ascertained afterwards in the state C_n, and then is subjected to a B-state test again, it *may* turn up in any of the states $B_1\ B_2\ \ldots$ individually unpredictable but with probability $P(C_n, B_k)$. In a similar fashion energy states and positional states of a particle are 'mutually incompatible', a fact of vital importance for the quantum mechanics of atoms.

Mutual incompatibility of two observable quantities, A and B refers to their probabilistic or statistical connection. It does not mean in general that two values A_k and B_j, cannot co-exist. In the example of orientation in a magnetic field it is impossible that a precessing magnetic particle has any fixed angle of orientation with respect to both fields B and C at the same time, both sets of states, B and C, being of the same general character. But there is no reason to deny that a particle, after having been tested by an energy meter and found to be in a certain energy state E_k, should not also have a certain position q at any time t, although states E and states (q, t) are incompatible. Incompatibility of A_k and B_j only means that a pair of values A_k and B_j cannot be *prepared* and cannot be *predicted*. Whether they can be *ascertained* simultaneously depends on whether one admits only 'direct' measurements, or allows reconstruction of data. More about this in Section 37. Bohr's idea that two such data can not *co-exist* is obviously of a metaphysical character. The empirical fact is only the impossibility of simultaneous preparation and prediction of a pair of data A_k and B_j.

Quite different from incompatibility of two data A_k and

B_j referring to two different observables A and B is the concept of mutual 'orthogonality' of two states of the same observable. An atom can be either in a state A_m or in the state A_n, but never in A_m and A_n at the same time. The probability relation between A_m and A_n is $P(A_m, A_n) = 0$, when m and n differ. A_m and A_n are *mutually exclusive*, also denoted as *mutually orthogonal* because of a certain geometrical analogy.

In all these considerations it is irrelevant whether the various values $A_1 A_2 \ldots$ of the observable A, the *eigenvalues* of A, represent a discrete set of values, or are distributed as a continuity or quasi-continuity. The eigenvalues of the energy E of a hydrogen atom are continuous for positive E-values but discrete for negative E's. Only for reasons of mathematical simplicity do we here consider the case of a discrete 'spectrum' of eigenvalues of an observable. Furthermore the same energy meter which is used for ascertaining values E also ascertains values of the observable $E^* = E^3$; it could therefore be denoted as an E^*-meter or as an E^3-meter.

One must be very careful with assigning the term 'state' and 'observable' within the quantum theory. Position q is not an observable with states q_n, but position at a certain time (q, t) is an observable. That is, the set $A_1 A_2 \ldots$ may signify space location at time t_A of the observable (q, t_A), and the set $B_1 B_2 \ldots$ may signify data of the observable (q, t_B). The B-measurement of (q, t_B) is not a repetition of the A-measurement (q, t_A). A-results are connected with B-results by definite probabilities $P(A_k, B_j)$. This shows that for non-conservative quantities such as position q, the instant of observation is essential. One also is wont to speak of energy states; this is not quite correct in the quantum sense of a 'well-defined' or 'reproducible state'. Just as energy is the sum of kinetic plus potential energy, so is momentum composed of kinetic plus potential momentum. The sum of the two is constant in space, as the sum of the two energy parts is constant in time. Strictly speaking, the combination (p, E) is an observable, constant in space and time, conjugate to the observable (q, t).

When speaking of 'states' as 'well-defined' only when they fit into the probability rules developed above, this is not a circular definition. On the contrary, the procedure of theoretical physics is always that of first being encouraged by certain spectacular examples to construct a certain schema expressed in mathematical symbols fitting those special examples. Later one extends the validity of the schema to new and properly chosen examples.

After having discussed 'states' of observable quantities, let us next treat of the *objects* to which the states pertain. Such an object is a hydrogen atom; it is capable of an infinite number of states with various internal and translational energies. Of more interest is a more 'restricted kind' of H-atom which is in one definite translational state, for example, with its centre of gravity at rest, and which still has a variety of internal states. One orthogonal set of internal states—we may denote it as the set A—is characterized by three quantum numbers nlm where n characterizes the energy, l the angular momentum, and m the component of the latter with respect to the fixed direction A of a magnetic field. (We disregard the spin.) Another set B of states of the same restricted kind of H-atom is characterized by quantum numbers nlm^*, where m^* refers to orientation with respect to a magnetic field of another direction B. Still another set of states is represented by various triples of 'parabolic' quantum numbers n_1 n_2 m for states oriented in an electric field; and so forth. Each of these orthogonal sets of states has the same infinite multiplicity, $M = \infty$. Next consider a still more restricted 'kind' of H-atom, for example, one with a definite energy E_4 with quantum number $n = 4$ once and for all. It is capable of a variety of states, one set being characterized by pairs of quantum numbers lm, another by lm^*, and so forth; each of these sets has the same multiplicity, $M = 16$ (since l can have values 0, 1, 2, 3, and m varies from l to $-l$). A still more restricted kind of H-atom is one with $n = 4$ and $l = 2$, say. Its orthogonal sets of states are characterized by quantum numbers m, or m^*; they all have the same multiplicity, $M = 5$. These examples bear out the theorem that the various orthogonal sets of states of a certain 'kind' of atom have *common multiplicity M*, yielding quadratic P-tables with M rows and M columns.

18. *Measurement*

Niels Bohr[2] in his essay, 'On the Notions of Causality and Complementarity', writes: 'In this situation we are faced with the necessity of a radical revision of the foundation for the description and explanation of physical phenomena. Here it must above all be recognized that, however far quantum effects transcend the scope of classical physical analysis, the account of the experimental arrangement must always be expressed in *common language* supplemented by the terminology of classical physics. This is a simple logical demand, since the word "experiment" can in essence only be used in referring to a situation where we can tell others what we have done and what we have learned.'

I have followed Bohr's 'logical demand' by using as testing instruments in the previous discussion only such arrangements, called 'filters' or 'separators' or 'meters' which are not appreciably influenced by the tested objects. They are *macrophysical instruments* which remain intact even when they throw the tested micro-object from its original state to one of the states characteristic of the 'meter'. For example, a double-refracting crystal is a 'separator' for two orthogonal states of polarization A_1 and A_2 whereas a Nicol prism is a 'filter' which passes A_1 and repels A_2. The Nicol itself, however, is supposed not to be affected by its own activity. Another example: What is commonly known as a yardstick, is a 'position separator' or 'position meter' which can be used over and over again to yield objective reproducible positions of a particle, relative to a fixed 0-point. Similarly, an energy meter is supposed to be so massive as to record energies of a particle without being affected itself; and so forth. The term 'measurement' and 'preparation of a state' (Margenau) ought to be reserved for a 'situation where we can tell others what we have done and what we have learned' (Bohr) in an objective confirmable sense. However, if 'states' are reproducible, then I cannot see why one should need a new *philosophy* of knowledge in which the word 'objective state' is eliminated. Quantum physics requires merely that we ought to be careful with applying the term 'state'. (q, t) is a state, and so is (p, E), but there is no combination (q, p)-state in the quantum sense. Similarly there are states of viscosity v of water, and states of twist w of a column of frozen water. Yet there are no reproducible combination (v, w)-states. It is true that the incompatibility of p and q, in contrast to the well-known incompatibility of v and w, has been a most important physical discovery. But if the vw-case does not call for a new philosophy or a new language, then the pq-case does not either. More about this in Chapter VIII.

Summary

(*a*) The purely formal content of Chapter IV may be condensed as follows. The elements of a class S of entities are supposed to be connected by positive fractions $P(S_l, S_j)$ less than unity, with only $P(S_l, S_l) = 1$. The elements of S can then be divided into subclasses $A, B, C, \ldots$ so that each subclass contains a complete set of mutually orthogonal elements with $P(A_k, A_{k'}) = \delta_{kk'}$. The P-fractions connecting elements of different subclasses are supposed to satisfy the unit sum rule, $\sum_j P(A_k, B_j) = 1$, for the rows of the P-matrices. The P's are to be symmetric, $P(A_k, B_j) = P(B_j, A_k)$. Hence, the unit sum rule holds also for the columns, $\sum_k P(A_k, B_j) = 1$, so that the P-matrices are 'unit magic squares'. The various subclasses $A, B, C, \ldots$ therefore have common multiplicity.

(*b*) The formal substructure becomes a physical theory by identifying the entities S with the quantal states of a mechanical system, and $P(A_k, B_j)$ with the fractional equality degree between the states A_k and B_j defined as a passing fraction in a test. The postulate of a two-way *symmetry* for transition probabilities corresponds to the classical reversibility of deterministic processes.

CHAPTER V

THERMODYNAMIC BACKGROUND

'Molecular disorder means not only excluding such ordered states as all molecules going parallel, but excluding the time reflection of every permitted state.'

SCHRÖDINGER

19. *Energy continuity versus equipartition*

It is a strange fact that the same quantum theory, which is notorious for its discontinuous 'quantum jumps', began with the need for restoring *continuity* to the thermal properties of bodies. Indeed, Max Planck found himself forced to introduce his quantum hypothesis, $\Delta E = h\nu$, for an oscillator of frequency ν when he studied the thermal energy content of many oscillators in a radiation field. Similar considerations apply to the thermal energy of a solid body, idealized as consisting of many oscillating particles, all with the same frequency ν resulting from a common binding force constant.

According to classical mechanics, the average energy per particle at temperature T should be $3kT$ (k = Boltzmann constant). This value, according to classical statistics, is to be independent of whether the particles are bound with high or low frequency, corresponding to a large or small binding constant. On the other hand, if the binding constant is increased to infinity so that the particles cannot move, their temperature energy should be *zero*. This result of classical statistical mechanics involves an unacceptable discontinuity, however. Suppose the binding constant and the corresponding frequency of vibration is gradually increased from small to large values; the average energy per particle supposedly remains $3kT$ even when the binding constant reaches enormous values. Only in the last moment, when the particles become entirely fixed, their average energy should suddenly

60

decrease from $3kT$ to zero—if classical theory is right. According to the general postulate of continuity of cause and effect, one would expect, and one actually finds, a *gradual* decrease, from $3kT$ to zero, of the thermal energy per particle with gradually increasing frequency v, rather than the same $3kT$ for all v's (equi-partition). So something must be wrong with classical theory.

Max Planck overcame the paradox of energy discontinuity for thermal energy averages by introducing the unclassical hypothesis that an oscillator of frequency v can change its energy E only in amounts $\Delta E = hv$, and thus can have energy contents which are multiples of the quantum hv above the smallest energy which is zero (or $\frac{1}{2}hv$). At low temperature, then, a tightly bound particle, or an electro-magnetic vibration of high frequency v, can climb to the high-energy level hv only at rare occasions, and will dwell mostly in the lowest level; thus it will exhibit a low average energy *gradually* approaching zero when v gradually approaches infinity. Planck's hypothesis was a stroke of genius; it was the one correct assumption, among many possible ones which would have restored continuity of the thermal energy. Yet it was an *ad hoc* hypothesis which could not have been deduced from the general and quantitatively non-committal postulate of energy continuity alone.

20. *Entropy continuity versus Gibbs paradox*

The postulate of entropy continuity which we are going to introduce in the following does not suffer from this defect. Without needing creative intuition it leads straight to those phenomena which are illustrated in Fig. 2c (p. 40). Entropy has always been defined with the help of 'filters' or semi-permeable diaphragms. Yet classical statistical mechanics suffers from a discontinuity pointed out first by W. Gibbs in the 1870's. In order to illustrate the *Gibbs entropy paradox*, imagine two equal quantities of the same kind of gas particles at the same temperatures in two equal volumes V separated by a wall.

The particles of the one gas may be in a state A, those of the other gas in the state B. For example, the particles may be silver atoms, and the atomic axis directions of gas A may point towards the north, those of gas B at an angle θ away from the north. (The axis directions may be conserved due to extreme dilution.) If we now remove the wall between the two volumes, either gas will spread over the combined volume $2V$. This diffusion process is connected with an increase of the entropy S and of the maximum isothermal work W which can be gained by letting the diffusion process take place in an isothermal reversible way, for example with the help of a semipermeable diaphragm which passes the one gas freely, but suffers a recoil pressure from the other gas. The maximum isothermal work δW and the entropy increase δS in the process of diffusion are

$$\left.\begin{array}{l} \delta W = 2NkT \cdot ln2 \\ \delta S = \ 2Nk \ \cdot ln2 \end{array}\right\} \text{ for } A \neq B \qquad (1)$$

where $2N$ is the total number of gas molecules in the two samples. On the other hand, if the two states A and B are *equal*, when there is no filter capable of separating them so that work cannot be gained from the diffusion process, in this case

$$\left.\begin{array}{l} \delta W = 0 \\ \delta S = 0 \end{array}\right\} \text{ for } A = B. \qquad (2)$$

Imagine now that the difference between A and B, for example the angular difference θ of the atomic axes, is gradually decreased and finally is made to disappear altogether. As long as there is a difference, however small, mixing of the two gases, according to classical ideas, yields the full maximum isothermal work and entropy increase (1). If finally the difference (for example the angle θ) is decreased from an extremely small finite value to exactly *zero*, then according to classical statistical thermodynamics, δW and δS decrease abruptly from the value (1) to *zero* as in (2). This is the *discontinuity paradox of Gibbs*, which classical theory cannot solve, although various answers have been proposed. For example, it is said that the paradoxical situation envisaged by Gibbs never occurs in nature since two gas species have a finite difference, or no difference at all, gas differences being determined by integral numbers, for example by the number of electrons in the molecules, etc. Therefore, *do not worry about the paradox*!

This seems to be disputing away the paradox rather than solving it. It is true that different species of gas atoms or molecules cannot be made equal in a gradual way. This argument fails, however, for different *states*, A and B, of the same kind of atoms. The considerations of Chapter III have shown that the general cause-effect continuity postulate requires the splitting effect of Fig. 2c, hence the existence of fractional equalities between two states, $A \sim B$, intermediate between $A = B$ (inseparability) and $A \neq B$ (separability). In the case $A = B$ the B-particles penetrate the A-filter, hence they do not exert pressure on it. In the case $A \neq B$ the B-particles produce pressure. In case of $A \sim B$ only the repelled B-particles exert pressure and contribute to the δS-value. When the state of the particles is gradually changed from $B = A$ to $B \neq A$ via $B \sim A$, the diffusion entropy δS gradually changes from the value 0 to $2Nkln2$.[1]

Silver atoms of spin pointing north cannot be totally separated from those of spin direction enclosing an angle θ with respect to the north. Their equality fraction is $P = \cos^2 (\frac{1}{2}\theta)$. In general when two gases, each consisting of N particles, have mutual equality fraction P, then their diffusion leads to an entropy increase δS intermediate between $\delta S = 0$ for $P = 1$ (total equality) and $\delta S = Nk \cdot 2 \cdot ln \, 2$ for $P = 0$ (total inequality) so as to solve the Gibbs paradox[1] by introducing entropy continuity when P is gradually changed from $P = 0$ to $P = 1$, as exemplified by decreasing the angle θ between the spin directions of two silver gases from 180 degs. to zero. Vice versa, when first postulating that the Gibbs paradox of entropy discontinuity (either $\delta S = 0$

or $= Nk \cdot 2 \cdot ln\ 2$) does not exist and is actually bridged by intermediate δS-values, one could have 'predicted that fractional separability must occur, together with such separation effects as depicted in Fig. 2c on p. 40.

21. *Entropy of diluted gases*

The failure of classical considerations to overcome the Gibbs paradox becomes obvious also in the entropy value of a single gas of N particles at absolute temperature T in a volume V with pressure $p = NkT/V$. Classical statistical mechanics leads to the following entropy S:

$$S(N, V, T) = Nk\{ln\ V + C \cdot ln\ T + \text{const}\}, \qquad (3)$$

where the *const* does not depend on V and T, and C is 3/2 for monatomic and 5/2 for diatomic molecules. This expression for S suffers from an insufficiency. Suppose there are seven equal gas samples under equal external conditions in separate volumes V. The total entropy is

$$\mathscr{S}_0 = 7 \cdot S(N, V, T).$$

Now remove the walls between the gas samples so that there now is one gas of $7N$ particles in the common volume $7V$, yielding the entropy

$$\mathscr{S} = S(7N, 7V, T).$$

According to (3) the entropy difference becomes

$$\mathscr{S} - \mathscr{S}_0 = \delta\mathscr{S} = 7\ Nk \cdot ln\ 7. \qquad (4)$$

Actually there is no entropy difference when like gases diffuse. The expression (3) must thus be changed in such a way as to yield $\delta\mathscr{S} = 0$ for like gases diffusing into one another—and intermediate expressions between $\delta\mathscr{S} = 0$ and the classical value (4) for gases of fractional equality quotients P, to overcome the Gibbs paradox. The correct entropy must be such that for seven *like* gases one has

$$\mathscr{S}_0 = 7 \cdot S(N, V, T) = S(7N, 7V, T) = \mathscr{S} \qquad (5)$$

instead of the value (4) which is correct only for the diffusion of seven different gases.

Now, since the dependence of S on V and T is thermodynamically assured, the only possible modification of (3) is to let *const* depend on N, and tentatively write

$$S(N, V, T) = Nk\{ln\, V + C \cdot ln\, T + f(N)\}, \qquad (5')$$

then determine $f(N)$ so that (5) is satisfied. Substitution of (5') into (5) yields for $f(N)$ the function

$$f(N) = -ln\, N + K$$

where K is a constant not depending on V, T, N, though K may depend on the sort of particles, for example on their mass. The correct expression for S now becomes

$$S(N, V, T) = Nk\{ln\, V + C \cdot ln\, T - ln\, N + K\} \qquad (6)$$

differing from the (wrong) classical expression (3) by the term $-Nk \cdot ln\, N$.

The constant K, however, which is related to the so-called Chemical Constant, cancels in entropy differences when the number of particles is conserved. For seven like gases one now obtains in agreement with (5) $\mathscr{S} = \mathscr{S}_0$. For seven entirely different gases, each diffusing from the original volume V into the common volume $7V$, one has to require $\mathscr{S} - \mathscr{S}_0 = 7 \cdot S(N, 7V, T) - 7 \cdot S(N, V, T)$, which according to (6) yields $7Nk \cdot ln\, 7$, which is correct, too. The case of diffusion of (two) fractionally equal gases yields (8) on p. 67.

The result that one must subtract from the classical entropy expression the term $Nk \cdot ln\, N$, which for large N is the same as $k \cdot ln\, (N!)$ (remember Stirling's formula), means that the classical probability of a configuration of N identical parts must be divided by the permutation factor $N!$. That is, *count the $N!$ permutations of N identical particles as only one single case.* Quantum theorists have come to this conclusion only in connection with rather late developments about the symmetry character of ψ-functions. We see here that the same result follows from the *non-quantal* purely thermodynamic requirement that diffusion of like gas samples must not increase the entropy. This consideration may contribute to the *demystification* of the quantum laws by their deduction from non-quantal postulates.

The same result may be obtained in an even simpler fashion. Imagine N molecules, all of the same kind and in the same state, situated somewhere within N equal adjacent volumes V_1 to V_N. Let the probability of this configuration be P_0 and its entropy $S_0 = k \, ln \, P_0$. Now imagine the walls removed so that the particles can interchange. According to classical viewpoints, the new probability, due to $N!$ permutations of N identical particles, now would be $N!$ times as large, and the entropy would be $S = S_0 + k \, ln \, (N!)$. In order to arrive at the correct result, $S = S_0$ (since no isothermal work can be gained from removing the walls) one has to omit the permutation factor $N!$. The foregoing considerations are from Schrödinger's *Statistical Thermodynamics*(2). They are independent of the typical quantum features (such as probability interference and pq-periodicity). At the same time it may be said that Boltzmann's equation, $S = k \, ln \, P$, is correct only when one chooses P so that Boltzmann's equation becomes correct.

When two gases, each containing N particles, diffuse from separate volumes V into the common volume $2V$, the entropy increase, $S - S_0$, is according to (6)

for equal gases: $S(2N, 2V, T) - 2 \, S(N, V, T) = 0$; (7a)
for different gases: $2 \, S(N, 2V, T) - 2 \, S(N, V, T)$
$$= 2kN \, ln \, 2. \qquad (7b)$$

How large is the diffusion entropy increase when the two gases have fractional equality P between 0 and 1 and go from the separate state to the mixed state (Fig. 3a)? In order to find out, we try to carry out an (adiabatic isothermal) unfusion

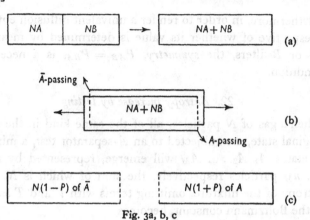

Fig. 3a, b, c

in two separate volumes again by inserting another volume $2V$ with an A-passing filter on its right, whereas the original volume is given an $\bar{A}$-passing filter on its left (Fig. 3b). Now we pull the two volumes $2V$ apart, ending up in Fig. 3c. The content of the two separate volumes is obtained by the following consideration. The A-passing filter on the right of the vessel not only passes the N particles A already present, but also splits the N particles B into NP_{BA} particles A remaining on its right, and $N(1 - P_{BA})$ particles $\bar{A}$ swept to the left and exerting pressure on the A-filter. The resulting two separate gases of Fig 3c then have total entropy (writing P for P_{BA}):

$$\mathscr{S} = S(N(1 - P), V, T) + S(N(1 + P), V, T).$$

Applying (6) to $\mathscr{S}$ and $\mathscr{S}_0$ we arrive at

$$\mathscr{S} - \mathscr{S}_0 = Nk[2\,ln\,2 - (1-P)\,ln(1-P) - (1+P)\,ln(1+P)]\ (8)$$

If the separation is carried out with a B- and $\bar{B}$-filter, the same value of $\mathscr{S} - \mathscr{S}_0$ would result, with P now standing for P_{AB}.

Irrespective of the relation between P_{AB} and P_{BA}, the entropy (8) *solves the Gibbs paradox*, because it yields a gradual change of $\mathscr{S} - \mathscr{S}_0$, from $Nk \cdot 2 \cdot ln\,2$ to zero, when the equality fraction P is gradually changed from 0 to 1.

Furthermore, in order to render a univalent diffusion entropy irrespective of whether its value is determined by means of A- or B-filters, the *symmetry*, $P_{AB} = P_{BA}$, is a necessary condition.

22. *Entropy increase by testing*

When a gas of N particles, all of the same kind in the same original state, are subjected to an A-separator test, a mixture of states $A_1 A_2 \ldots A_M$ will emerge, represented by $n_1 n_2 \ldots n_M$ particles respectively, the sum of which is N. The entropy of the mixture, omitting terms with V and T as well as the Boltzmann constant, becomes

$$S_A = -\sum_k n_k \ln n_k ,$$

larger, that is less negative, than the original entropy

$$S_0 = -N \ln N.$$

If we now subject the A-mixture to a B-separator test, the particles will jump to the new states $B_1 B_2 \ldots B_M$ resulting in new occupation numbers

$$m_j = \sum_k n_k \cdot P(A_k, B_j)$$

yielding the entropy

$$S_B = -\sum_j m_j \ln m_j .$$

It has been proved by von Neumann[3] on grounds of the unit sum rule for the rows and columns of the P-matrices that S_B is larger than S_A. Similarly, S_C resulting from a C-separator test is larger than S_B, and so forth. Test after test produces an increased entropy:

$$S_0 < S_A < S_B < S_C \ldots \tag{9}$$

With great probability S steadily increases toward a maximum value which belongs to *equipartition*, $n_1 = n_2 = \ldots n_M$ over the M-states.

We are faced here with an odd difference between classical and quantum physics. In classical theory, in order to obtain the Second Law, the reversibility of mechanical processes must be forbidden, according to the motto of this chapter. Yet the classical reversibility is the model of the two-way symmetry of the probabilities which leads, via the unit sum quality of the P-matrices, to the irreversibility of the entropy increase obtained from testing. *Reversibility* (symmetry) of individual processes entails *irreversibility* of averages. This is analogous to the former results that *causality* for averages entails *acausality* for individual events, and *continuity* of averages entails *discontinuity* of individual processes.

One must not forget, however, that statistical results such as (9) hold only in the overwhelming majority of cases; actually there are individual fluctuations away from the average. Thus although the entropy of a gas mixture will most probably increase by virtue of the next separation test, it may actually decrease, and under a long series of tests, the entropy will fluctuate up and down near the entropy maximum, as illustrated by the up and down staircase curve for S as a function of time t.

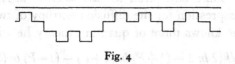

Fig. 4

This curve (Fig. 4) was first drawn by P. and T. Ehrenfest in order to illustrate the entropy of a gas according to *classical statistical mechanics*; the various staircase levels were to signify S-values obtained in a series of inspections of the gas from outside. Yet, whether one assumes classical inspection, or quantum infringement by testing, the fluctuating S-curve is of an overall *symmetry* with respect to the positive and negative time direction.

23. *Entropy and information*

It is taken for an almost self-evident truth that entropy and lack of information go parallel. Yet this is debatable since information is subjective, changing from person to person, whereas entropy is an objective physical quantity dominating thermodynamic phenomena, whether a person looks on and gains information or not. Let us consider a typical case.

Two equal volumes may contain N silver atoms each. Those in the one volume may point with their spin axes toward the north, those in the other at an angle θ away from the north. The wall between the two volumes is now removed so that spontaneous diffusion takes place. How large is the increase ΔS of the entropy S during the diffusion process? It may be argued that different observers will give different answers depending on their general knowledge of theory, their special information about the gases, and their technical ability of gaining isothermal work from the process.

A person who does not know that silver atoms have spin will judge the two gases to be *equal*; to him ΔS is *zero*. When he learns of the difference of the spin orientations he will change his mind and plead for $\Delta S = 2kN \cdot ln\,N$, which is the usual expression for the diffusion entropy of two *different* gases. If he knows more of quantum theory he will vote for

$$\Delta S = Nk\{2\,ln\,2 - (1+P)\,ln\,(1+P) - (1-P)\,ln\,(1-P)\} \quad (10)$$

where P is the 'equality fraction' of the two gases, in the present case $P = \cos^2(\tfrac{1}{2}\theta)$ which yields $\Delta S = 0$ for $\theta = 0$ and $\Delta S = 2kN \cdot ln\,N$ for $\theta = 180$ deg. as limiting cases of equality and inequality.

Shall we concede, then, that each of the three observers is right according to the *information* available to him, and that still other answers may be given by future physicists with more information? My own view is that there is but one true value ΔS independent of personal knowledge—and this on the grounds of the very quantum theory which allegedly involves knowledge by observing subjects. Quantum theory characterizes the 'state' of each particle by a few quantum numbers. Even the velocity is quantized in a finite volume. And although we do not yet know all the quantum features of nuclear particles, we can trust that their states are determined by a *finite* number of state characters. If an observer does not know of them, it is his fault and does not affect the one, true, and ultimate value of ΔS.

Whereas classical theory could claim with a certain right that the entropy has different values depending on the degree of knowledge and the methods of separation possessed by the observer, the quantum theory sets an upper limit to the possible separability of states and thus yields an objective value for the entropy change, irrespective of the knowledge which an observer has or thinks he has. Whether he is ignorant of the difference of the two gases and assumes $\Delta S = 0$, or whether he overestimates the difference between the two gases and pleads for $\Delta S = 2kN \ln N$, the actual ΔS is given by the expression (10) above.

The view that thermal quantities such as entropy differences have objective values is opposed by certain physicists today. For example, you have always thought that irreversible processes, such as the levelling of temperature differences, go on without anyone looking or knowing of them. This is not so: 'Imperfect knowledge defines the thermodynamic behavior of a system . . . and irreversibility arises from the incomplete knowledge of the initial conditions.' says L. Rosenfeld[4]. One here wonders whether the irreversible processes preceding the birth of a supernova also 'arise from the incomplete knowledge of the initial conditions'. Overlooking the difference between what *is* and what we or others *know* about it will always unleash confusion, as it has done in the interpretation of quantum theory to a great extent (see Chapter VIII). But applying 'incomplete knowledge' to thermal processes creates absolute Darkness at Noon. Rosenfeld may perhaps be excused by Schrödinger's dictum: 'What people mean when they say something else than what they mean, is hard to guess.' Yet on another occasion Rosenfeld startles us again by commenting on a 'connection between the statistical aspects of the quantum theory and thermodynamics. There is of course [!] no such connection.'([5]) This would have come as a distinct surprise to Max Planck who, in 1900, established the quantum theory precisely on statistical thermodynamic grounds.

Another example of drawing knowledge into the discussion of thermal quantities is due to Heisenberg([6]): 'When we describe the temperature of a piece of matter we define it objectively with the help of a thermometer. But when we try to define what the temperature of an atom could mean we are, even in classical physics, in a much more difficult position. Actually we cannot correlate this concept with a well-defined property of the atom but have to connect it, at least partly, with our insufficient knowledge of it.' In contrast to Rosenfeld, Heisenberg at least concedes that the thermal properties of a piece of matter are objective, and that knowledge 'at least partly' (how partly?) matters only for atoms. To others it would seem that 'the temperature of an atom' is a contradiction in terms, and that temperature, as measured with a thermometer, has never anything to do with insufficient knowledge. When a student answered the question 'Define the Second Law' by writing: 'The confusion of the Universe is steadily increasing', he apparently confused the Universe with the quantum language.

24. *Entropy and the direction of time*

After a century of statistical thermodynamics there is still lack of unanimity concerning the question whether the trend toward higher S-values can be used consistently for defining the direction of time, in spite of the two-way symmetry of processes in classical mechanics, and the two-way symmetry of probabilities in quantum theory. Let us therefore consider an example. When the temperature of a body is found to be high on one side and low at the other side, and at another time the temperature is even throughout, and the body was heat-insulated during the time interval, then we normally assign the earlier time instant to the hot-cold state and the later instant to the even temperature state. The time direction can here be found from the entropy direction, both being positive simultaneously. The question is, however, whether this parallelism holds consistently and universally, or at least in the great majority of cases.

Let us first consider a finite volume containing a finite number of molecules, heat-insulated all the time and checked as to the molecular locations and velocities in a series of observations stretching over an infinite time span. The corresponding distribution probabilities P and entropies, $S = k \ln P$, may be recorded on a film strip. (We assume that the inspection does not affect the distribution, and that the latter can indeed be ascertained, a highly idealized case of classical theory.) The recorded entropy curve will show many ups and downs, most of the time residing near the maximum of S and only rarely descending to low S-values, as shown by the famous staircase curve of P. and T. Ehrenfest (Fig. 4). When the film is projected as a moving picture, one cannot tell whether it is run off forward or backward since, over a very long time interval, the S-curve has an overall *two-way symmetry* in time. Time proceeds in one direction, but the entropy runs up and down. Boltzmann's original plea for a probability favored one-sidedness of the entropy development is indeed contested by the Lohschmid inversion argument: Any motion of the molecules leading to an S-increase will, by inversion of the molecular motions, lead to a corresponding S-decrease; and there is no mechanical preference of the one motion over the opposite one. There also is the Poincaré-Zermelo recurrence argument: If one only waits long enough, he will find a repetition or near-repetition of any given section of the S-curve. But if there is a recurrence of an upward S-curve section, it entails a downward section between the two up-

ward sections. Hence there can be no systematic parallelism
between entropy and time's arrow. The same negative result
is obtained from quantum physics where 'inspection' is replaced by 'testing'. Although each test with great probability
leads to a higher S-value, the actual S-curve obtained from
an unlimited number of tests yields an S-curve with up and
down fluctuations, similar to the Ehrenfest curve, likewise
showing two-way symmetry in time.

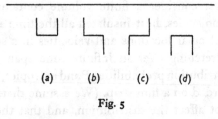

(a)   (b)   (c)   (d)

Fig. 5

In order to understand that nevertheless low entropy values
are much rarer than high ones, Ehrenfest considered the
following three-step sections (Fig. 5) recurring along the
S-curve. In case (a) a certain low S-value is followed as well
as preceded by a higher S; this is the most frequent case.
Less probable are cases (b) and (c), where the same low S-
value is passed, in (b) on a downward, and in (c) on an upward,
three-step section. (b) and (c) have the same probability, much
less than that of case (a). The most improbable case is (d)
where the same low S-value is preceded as well as followed
by a still lower S-value. But (a) as well as (d) are $\pm$ t-sym-
metric, and the same holds for (b) and (c) together. In view
of inversions and recurrences, no general parallelism of
entropy and time direction prevails, in classical as well as in
quantum theory, for a finite system.([7])

In an infinite system, however, one shall have to wait an
infinite time until an inversion or a recurrence of an S-curve
section will occur. For example, an originally space-condensed
group of molecules may expand into infinite space forever, thus

systematically increase the entropy with increasing time. The same holds for an expanding universe of stars, supposing that it expands into infinite space.

In ordinary life and in thermodynamics we deal with finite systems which deliberately are brought into a state of high entropy by branching them off from the rest of the universe. Here we have indeed, at least for a limited time span, a parallelism of time and entropy—which does not prove any such parallelism in general, that is for the whole Ehrenfest curve. We happen to live in a very improbable state of the world which develops into a more probable state. But if ours were not a very improbable world, there would not be any '*we*' to speculate about it.

Summary

A solution of the Gibbs paradox of entropy discontinuity is reached by introducing the concept of a fractional degree of equality between states, operationally defined already in Chapter II. The reduction of $N!$ permutations of N identical particles to unity is as fundamental for quantum theory as the two-way symmetry of the transition probabilities. It precedes the interference of probabilities and the pq-periodicity, to be dealt only in later chapters. The two-way symmetry of classical mechanics is the chief barrier against a derivation of the Second Law on a classical basis. In the new mechanics, however, reversibility (= two-way symmetry) of the individual probabilities is the prerequisite for the irreversibility of averages. Nevertheless, neither classical nor quantum theory permits a *general* entropic definition of the direction of time. The thesis that entropy goes parallel with lack of information suffers from a confusion of information, that is actual information, with theoretically possible information which is not information.

CHAPTER VI

INTERFERENCE OF PROBABILITIES

'It is my conviction that we can discover by pure mathematical construction the concepts and the laws which furnish the key to the understanding of natural phenomena.'

EINSTEIN

25. *The metric of geometrical structures*

This chapter deals with the interference or superposition of probabilities, usually regarded as a sovereign quantum law, as an oddity of nature which cannot be explained on the grounds of more elementary postulates. It will be shown here that probability interference is a mathematical necessity under the *non-quantal* postulate that the transition probabilities from state to state are interdependent under a general law, rather than presenting a chaos of independent entities, the only general law connecting unit sum square tables being that of unitary transformation; it is formally identical with the interference law of quantum mechanics. The latter is not only a fact; it is a necessary fact.

Someone engaged in folding paper triangles finds that the three altitudes in every triangle intersect in one point, and that the same holds for the three angular bisectors, the three medians, and the three perpendiculars on the side-centers. He concludes that he has discovered a new general principle of 'unity in triplicity' which is confirmed not only in geometry but also in music, architecture, political institutions, and so forth. A mathematician will tell him, however, that his geometrical unity in triplicity is not a separate 'principle' in its own right, that it is a consequence of the Euclidian axioms, and can thus be understood as a geometrical necessity, instead of being accepted as a cabbalistic oddity with deep philosophical implications. Similarly, the empirical fact that the statistical behaviour of particles is wavelike may perhaps be denoted by the comprehensive term 'wave-particle duality'. However, when physicists after half a century of quantum physics still believe that duality, together with a complementarity of particle- and wave-features, reveals an independent fundamental 'principle' of nature, an immanent trait of the microcosm with ramifications in other domains of human knowledge[1], then it is time to pause and reflect whether this duality might not be reducible to simple, almost self-evident physical ground axioms, so that we can recognize it as a necessity rather than an oddity. I think indeed that simple and elementary postulates can be established which explain *why* the statistical manifestations of particles should obey a law of 'interference of probabilities' *via* a complex-imaginary amplitude called ψ, and why $E = h\nu$ and other quantum rules prevail, that is why the co-ordinates and momenta of particles should be in a periodic wavelike relation. Such a reduction from the strange to the natural is all the more desirable as the alleged mysterious character of the quantum rules seems generally to be taken for granted, as underlined, for example, when P. Bridgman[2] contends: 'When we go far enough in the direction of the very small, quantum theory says that our forms of thought fail, so that it is questionable whether we can properly think at all,' and when Heisenberg[3] declares that the quantity ψ is 'abstract and incomprehensible . . . so to speak containing no physics at all'. In order to dispel the aura of incomprehensibility surrounding the esoteric quantity ψ let us first consider a geometrical analogy.

Suppose an infinite number of cards are found to be inscribed with positive numbers such as $L_{AB} = 5.23$ and $L_{BC} = 4.78$, and so forth, among them $L_{AA} = 0$, etc. The L's are symmetric, $L_{AB} = L_{BA}$. We suspect that the writer of the cards has used a certain method of number assignment so that L_{AC} is either determined, or at least restricted in its value by the values of L_{AB} and L_{BC}, and that the same 'triangular' relation holds for any other triple of cards. That is, suppose $F(L_{AB}, L_{BC}, L_{CA}) = 0$ or shortly written $f(A, B, C) = 0$ describes the supposed interdependence. The same formula shall hold also for any other triple of letters without yielding a contradiction. In particular, from $f(A, B, C) = 0$ and $f(A, B, D) = 0$ and $f(A, C, D) = 0$ shall follow $f(B, C, D) = 0$ as a consequence through elimination of A. In other words, the interdependence law between the L's is to be *symmetric* in its three letters as well as *transitive*, or to have a group character. We call this the condition of *generality* of the interdependence law. These restrictions alone are not sufficient to determine the method applied by the card writer. But when it is found in addition that the L's satisfy the inequality

$$| L_{AB} - L_{BC} | \leqslant L_{AC} \leqslant (L_{AB} + L_{BC}) \qquad (1a)$$

this reveals the L-law is a univalent way, best illustrated by a geometrical picture: Imagine the letters $A, B, C, \ldots$ to represent points in Euclidian space, one- or two- or M-dimensional, and the L's to be the distances between the points.

The L-relation may then be formulated also in the following manner: Associate with every positive $L_{AB} = L_{BA}$ two opposite *vectors*, $\omega_{AB} = -\omega_{BA}$ of magnitude $|\omega| = L$ and give them such directions that they satisfy the triangular addition theorem

$$\omega_{AB} + \omega_{BC} = \omega_{AC} \text{ with } \omega_{AB} = -\omega_{BA}, \text{ hence } \omega_{AA} = 0 \quad (1b)$$

or in a more symmetric form

$$\omega_{AB} + \omega_{BC} + \omega_{CA} = \omega_{AB} + \omega_{BA} = \omega_{AA} = 0 \qquad (1c)$$

The vector composition theorem is symmetric and transitive, as required for a general interdependence theorem, and it looks much simpler than the direct connection between the positive quantities L themselves where one must distinguish between various cases, as follows:

(a) If the points $A, B, C, \ldots$ are distributed in one dimension on a straight line (or on a circle) then two lengths L_{AB} and L_{BC} determine L_{AC} only in a bivalent fashion since C may be located on either side of B and A. However, the six connections L between four points A, B, C, D are in a fixed relation so that five L's uniquely determine the sixth L, by a general mathematical procedure which is symmetric in the letters and has group character, that is, can be continued to any other set of points. The L's are in a *tetragonal relation*, one-dimensional structures being supported by four points.

(b) If the points are on a plane (or a spherical surface) then one has a *pentagonal* relation between the L's. Five points are connected by ten distances L. When nine of them are given, they determine the tenth L by a unique correlation law which is symmetric in the letters A, B, C, D, E and holds for any five letters.

(c) In three dimensions there is a hexagonal L-relation law. Six points are connected by fifteen L's. Fourteen of them uniquely determine the fifteenth.

And so forth for higher dimensional spaces, flat or of constant curvature. Although the L-correlation laws become more and more involved with higher dimensions, the simple vector law (1b) holds in any number of dimensions.

Notice that the condition $L_{AB} = L_{BA}$ does not contain $L_{AA} = 0$ as a special case, in contrast to $\omega_{AB} = -\omega_{BA}$ implying $\omega_{AA} = 0$. This is connected with the fact that there can be no *unique* general relation between three L's themselves, and that one has to resort to *auxiliary quantities* ω not uniquely determined by the corresponding L's.

26. *The metric of probability amplitudes*

We now proceed to the more difficult mathematical problem of constructing an interdependence theorem between the P-matrices which determines, or at least restricts, the elements of the matrix (P_{AC}) when those of (P_{AB}) and (P_{BC}) are given. The desired 'triangular' connection theorem should of course be *general*, which is to mean:

(1) If we denote the interdependence symbolically as $F(P_{AB}, P_{BC}, P_{CA}) = 0$, or still shorter as $f(A, B, C) = 0$, then the same connection ought to hold also for any permutation of the letters A, B, C; that is the theorem ought to be *symmetric* in the letters.

(2) The interdependence ought to hold for *all* letter combinations so that from $f(A, B, C), f(A, B, D), f(A, C, D) = 0$

79

should *follow* $f(B, C, D) = 0$ by elimination of A. This is requiring *transitivity* of the theorem.

The conditions of symmetry and transitivity together will be denoted as those of *generality*.

The first question is whether a general law of P-matrix interdependence is possible at all. This is answered in the positive by just constructing one (see below). The next question is whether the constructed example is the *only possible* interdependence law between P-matrices. Uniqueness can also be shown (see Appendix 1). Apart from the unit sum square quality of the P-matrices, their positive elements must satisfy the condition of two-way symmetry

$$P(A_k, B_j) = P(B_j, A_k) \text{ together with } P(A_k, A_{k'}) = \delta_{kk'}. \quad (2)$$

We now begin the task of constructing a P-matrix interdependence theorem. In order to indicate that the members of the P-tables are positive and between 0 and 1, we introduce the notation α^2 for P where α is a real positive or negative quantity between -1 and $+1$, hence $\alpha = \pm \sqrt{P}$. The matrix (P_{AB}) then looks as follows:

$$\begin{pmatrix} \alpha^2(A_1, B_1) & \alpha^2(A_1, B_2) & \ldots \\ \alpha^2(A_2, B_1) & \alpha^2(A_2, B_2) & \ldots \\ \ldots & \ldots & \ldots \end{pmatrix} = (\alpha^2{}_{AB}). \quad (2a)$$

Each of the M rows and M columns adds up to unity.

The notation α^2 for P has been introduced because it immediately suggests a geometrical interpretation: Let the letters $A_1 A_2 \ldots A_M$ represent M perpendicular ($=$ orthogonal) axes in M-dimensional space, and let the letters $B_1 B_2 \ldots B_M$ stand for another orthogonal axes system. The quantity $\alpha(A_k, B_j)$ may be interpreted as the cosine between the two axes A_k and B_j. The squares of the cosines in the matrix (2a) then add up to unity in each row and each column. This geometrical interpretation leads to a solution of our problem of finding a *general* (symmetric and transitive) interdependence theorem connecting unit sum square matrices:

Associate with every positive P a positive or negative quantity $\alpha = \pm\sqrt{P}$ and connect the α's by the formula for cosines:

$$\alpha(A_k, C_n) = \sum_j \alpha(A_k, B_j) \cdot \alpha(B_j, C_n) \text{ with } P = \alpha^2, \quad (2b)$$

known as *orthogonal transformation*. It solves the problem of finding a general self-consistent interdependence theorem between the P-matrices, leaving their unit sum quality *invariant*. One can write the relation (2b) in the more condensed notation as matrix multiplication:

$$(\alpha_{AC}) = (\alpha_{AB}) \times (\alpha_{BC}) \text{ with } (\alpha_{AA}) = (1) \quad (2c)$$

or in the symmetric form:

$$(\alpha_{AB}) \times (\alpha_{BC}) \times (\alpha_{CA}) = (\alpha_{AB}) \times (\alpha_{BA}) = (\alpha_{AA}) = (1) \ (2d)$$

in analogy to the vector addition (1b) (1c) which solved the problem of interdependence between the quantities L before. The sign, $+$ or $-$, of an individual $\alpha = \pm\sqrt{P}$ is left open. It is restricted, however, by the mutual interdependence (2, b, c) between the α's.

There is an important generalization of the schema of orthogonal transformation which also safeguards the unit sum quality of the P-tables to be invariant. It is known as *unitary transformation*. Instead of real positive or negative quantities α one can introduce complex quantities ψ and write $P = |\psi|^2$ so that, with ϕ being a phase:

$$\psi = \sqrt{P}\, e^{i\phi}, \text{ vice versa } P = |\psi|^2 \quad (3a)$$

whereby ψ is to be 'hermitian', that is

$$\psi(A_k, B_j) = \psi^*(B_j, A_k) . \quad (3b)$$

The asterisk signifies the complex conjugate. The quantities ψ are to satisfy the same relation which for the α's was given in (2b). In the abbreviated matrix notation they are connected by the formula

$$(\psi_{AC}) = (\psi_{AB}) \times (\psi_{BC}) \text{ with } (\psi_{AA}) = (1) . \quad (4)$$

Each complex quantity ψ is afflicted with a phase factor $e^{i\phi}$ which is not determined by the corresponding quantity $P = |\psi|^2$. The phases ϕ are restricted by the mutual interdependence (4). Restriction to the values $\phi = 0$ deg. and 180 deg. yields the special case of orthogonal transformation. The phase has no significance for an individual probability P.

Since a complex quantity ψ can be represented geometrically as a *vector in a plane* with a real and an imaginary axis, the theorem of unitary transformation (3a, 3b, 4) stipulates that the square roots of the probabilities P can be laid out as vectors in a plane so as to represent a kind of two-dimensional framework, dominated by the matrix product theorem (4), in analogy to the vector composition law (1c). All this shows that unitary transformation is a possible solution of the problem to find an interdependence law connecting P-matrices.

It is important to make sure, however, that 'interference' is the only possible *general* law connecting the probability matrices so that one can say that it is a necessity, that *Nature has no other choice* to avoid chaos. The uniqueness proof is given in Appendix 1. It rests on trying out all possible general, that is symmetric and transitive, connections between triples of quantities ω_{AB}, ω_{BC} and ω_{CA} up to the third order (there are only a few possibilities) because in any combination of higher than third order the three quantities occur individually in higher than first order which leads, in the test for transitivity, to quadratic, cubic, etc., equations with bivalent or multivalent solutions which are irrational or transcendental and cannot pass the test of transitivity. The question of *why do the probabilities interfere* can thus be answered: They have no other choice if they 'want' to obey a general interdependence law at all. The special question why Nature is not satisfied with orthogonal transformation involving real quantities α but resorts to complex quantities ψ will be answered in Chapter VII and Appendix 3.

Notice that any one single transition probability $P(A_k, B_j)$ is as little 'associated with' a wavelike amplitude and phase, as an individual rod of length L_{AB} is associated with a vector direction. It is true that one *can* lay P (or L) in a certain direction and thereby fit the corresponding vectors ψ (or ω) into the mathematical schema of unitary transformation (or vector addition). The idea, however, that an *electron* travelling through space is a 'wavicle' having phase and amplitude is utterly mistaken.

The following remarks of a mathematical nature may be added. It is characteristic of the unitary transformation (4) that it has a very simple and general structure, even when written out in detail as

$$\psi(A_k, C_n) = \sum_j \psi(A_k, B_j) \cdot \psi(B_j, C_n). \qquad (5a)$$

Here A, B, C are *any* three observables, and the subscripts k, j, n indicate *any* of their eigenvalues. The simple structure of this general theorem is spoiled, however, by the customary way of specializing it. First, B is given the special meaning of energy with eigenvalues E_j; and C is taken as the positional co-ordinate q, still leaving A unspecified, so that (5a) reads

$$\psi(A_k, q_n) = \sum_j \psi(A_k, E_j) \cdot \psi(E_j, q_n). \qquad (5b)$$

Then, because q has a continuity of values, the subscript n is omitted. Furthermore, reference to the state A_k is omitted entirely, and (5b) is written as

$$\psi(q) = \sum_j a_j \psi_j(q) \qquad (5c)$$

as an *expansion* of the 'state function' $\psi(q)$ in terms of, or as a superposition of the eigenfunctions of the energy, $\psi_j(q)$. (For q one also can take the position in space and time, q, t.) This way of writing may be convenient for practical purposes as a shorthand notation. But many textbooks take it as their starting-point and as a 'principle of superposition' in an axiomatic fashion. However, the student then has great difficulty in finding out that this 'ugly' special sum is but a disfigured form of the 'beautiful' and general theorem (5b) which, moreover, can be understood as a consequence of non-quantal postulates, as shown above. It is true that the historical development began with the form (5c) in Schrödinger's wave mechanics. It would seem, however, that the adept today has a right to be introduced to the quantum theory on the straight and beautiful highway opened by the consistent and general notation (5b) rather than struggle along on the erratic and unsystematic byways of the historical development, so that only after years of practice he learns of the simple and general underlying structure in an understandable rather than mnemotechnical way.

27. *The metric of probabilities*

The formal analogy between the ω-vector addition law (1a) and the ψ-multiplication law (3b) goes even further. As mentioned before, when five points A, B, C, D, E, in a *plane* are connected by ten distances L, or ten square distances P, nine P's uniquely determine the tenth P. Similarly, since the ψ's are complex numbers representing vectors in a *plane*, it can be shown that the triangular ψ-matrix law (4) is equivalent to a direct relation law between ten P-tables which connect any five orthogonal sets of states A, B, C, D, E so that nine unit magic squares uniquely determine the tenth.

This mathematical result has interesting implications. From the very beginning plane geometry has consisted mainly in exploiting special cases of the unique relationship between ten lengths in plane structures connecting five points. Only around the year 1900 did the vector notation come into general usage without, however, contributing any new insight into geometrical relations. The development of probability theory in physics has taken the opposite path. Here one first has discovered, after a long path of trial and error, the triangular product law (4) for matrices of complex quantities ψ, that is probability vectors in a plane. The same ψ-law leads to a direct law between ten P-matrices (as in (b) at the top of p. 79). The moral is that the interference law of probability amplitudes ψ is no more and no less essential for the interconnection between transition probabilities P than the belated introduction of vectors has been for geometry in a plane. Of course, nobody will overlook the great practical value of the ψ-law for finding quantitative relations between probabilities P in those cases where the phase relations of the corresponding amplitudes ψ happen to be known from independent sources. [They are known in particular for the standard function $\psi(q, p)$, Chapter VII.] One must not forget, however, that the phases, that is the directional angles, occurring in a single matrix (ψ_{AB}) are irrelevant for the probabilities in the single matrix (P_{AB})—just as the direction of a single rod is irrelevant for its length. Only when one wishes to describe relations between at least three P-tables (P_{AB}), $P_{BC})$, (P_{AC}), or between three sides of a triangle, then and only then do the vector laws become operative.

If one should restrict the quantities ψ to the *real* values $\psi = \pm \sqrt{P}$, that is if one allows only two opposite directions along a line, 'unitary transformation' would degenerate to 'orthogonal transformation'. The resulting P-tables would still be unit magic squares; but they would belong to a restricted class of such squares, too narrow for the requirements of quantum mechanics. The reason for admitting complex probability amplitudes, that is vector directions in a *plane*, rather than only along a line, will become clear in Chapter VII on the grounds of an invariance requirement of mechanics.

Probabilities can be used for calculating *mean values*. Let $S_1 S_2 \ldots$ be the M values which the observable S pertaining to a certain atom may possibly display under an S-meter test. Suppose now that the atom is first (or last, see below) subjected to an A-measurement showing the value A_k. Then it undergoes an S-measurement. The individual outcome is uncertain, but the mean value of S in case of many repetitions of the experiment, starting always from A_k is

$$S_{mean} = \sum_j P(A_k, S_j) \cdot S_j . \tag{6}$$

We may denote it as the mean value of S *from*, or contained *in*, the state A_k. The same result can also be written in the form

$$S(A_k, A_k) = \sum_j \psi(A_k, S_j) \cdot S_j \cdot \psi(S_j, A_k) , \qquad (6')$$

or still more abbreviated

$$S_{kk} = \sum_j \psi_{kj} S_j \psi_{jk} . \qquad (6'')$$

It may also be denoted as the mean value of S *in transition* from A_k to itself, or simply 'the transition value of S from A_k to A_k. Mathematically the last formula characterizes the observable S as a *tensor* in the M-dimensional space of unitary transformation, with various states as axes, S_j as principal values of the tensor S, and S_{kk} as tensor components with respect to those axes k (states k) in which another observable A has its principal or eigenvalues A_k. The tensor S also has 'mixed components' with respect to the axes k, defined as

$$S_{kk'} = \sum_j \psi_{kj} S_j \psi_{jk'} \qquad (7)$$

and more general

$$S_{kk'} = \sum_m \sum_{m'} \psi_{km} S_{mm'} \psi_{m'k'} \qquad (8)$$

with the special case $S_{jj} = S_j$ and $S_{jj'} = 0$. The probability amplitude ψ plays the part of the 'tensor unity'. *Quantum mechanical calculations then reduce mathematically to the transformation theory of tensors in unitary space.* One may introduce the tensor quality of observables as a separate axiom of quantum mechanics. However, the relation (6), when written in the form (6') (6''), implies this tensor quality already without extra assumptions. More about tensor calculus may be found in Appendix 2. On the other hand, after unitary transformation has been established as the connection between the most common observables (energy-momentum, space-time location, etc.), only those observables and their 'states' are further admitted to the treatment of quantum mechanics which fit into the schema developed here. The state of health of a body certainly is not an 'observable state' in the quantum sense and does not submit to unitary transformation.

It is often regarded as perplexing that the probabilities of atomic processes obey a law of interference rather than one of addition. However, addition would not even serve under ordinary circumstances. A cat can easily jump from point A_k to various points B_1 B_2 ... about ten feet away, and from there to a point C_n ten feet away from the points B. But it has no chance of jumping directly the twenty feet from A_k to C_n. It would be quite wrong therefore to calculate the latter (vanishing) probability as

$$P(A_k, C_n) = \sum_j P(A_k, B_j) P(B_j, C_n) \quad \text{(wrong!)}$$

As a general law this formula is likewise unfit since it would have to include the special case

$$P(A_k, A_{k'}) = \sum_j P(A_k, B_j) P(B_j, A_{k'}) \quad \text{(wrong!)}$$

with zero on the left for $k \neq k'$ and with a sum of positive products on the right. On the other hand, the fact that the probabilities of atomic processes obey as simple and general a law as that of interference signifies that we have arrived here at a very fundamental level of theoretical analysis with little chance of finding a still deeper subatomic level—in contrast to certain theorists who would like to search for a deeper stratum where our present statistical laws might prove to stem from causal ones. The search for hidden causes is an attempt of reducing simple laws of nature to complicated *ad hoc* invented ones (refer to Section 43), contrary to every standard of scientific method.

According to the consideration above, the relation between probabilities *via* the simple law of ψ-interference does not have to rely on any 'principle' of wave-particle duality, or on a hypothesis that particles are guided by wavelike laws to their proper places. Interference of probabilities, far from being perplexing, is the one and only possible way to connect unit sum squares by a general interdependence rule.

As seen before, an individual ψ-vector direction or phase does not have physical significance of its own. The complex quantities ψ with their directions in a plane merely serve to establish a mathematical relation between various transition probabilities plotted in tables (P_{AB}), etc., and observable in tests with macroscopic instruments, shortly called A-meters, B-meters, and so forth. This ought to dispose of the idea that a ψ-function or ψ-wave, which represents one row in a ψ-matrix, has any direct physical substantiality as a 'pilot wave' or 'quantum potential' which 'guides' events (de Broglie, Bohm, Vigier), or that it gradually expands and occasionally shrinks suddenly (Heisenberg, Section 39). A ψ-wave does not guide actual events any more than a mortality table guides actual mortalities, and it shrinks no more than a mortality table shrinks when an actual death occurs. Material waves were in order forty years ago; in fact they inspired the founders of wave mechanics to establish their marvellous method of calculating atomic data. Today the same ideas, after being imprinted into the brains of a whole generation, obscure rather than clarify the significance of all that quantum theory stands for, which is: to provide formulas, tables, and other rules of correlation between events, in particular between probabilities of transition. These rules ought to, and can be understood as *necessities* rather than as oddities under simple ground postulates.

There have been several ingenuous attempts to modify or generalize quantum theory by the introduction of new forms of matrix algebra in which the law of association, and that of commutation is dropped on a higher level, (refer to P. Jordan([4])). The question is always: do these mathematical developments represent new possibilities of correlating unit magic square tables? If not, then I cannot see too much chance that they will lead to progress in quantum physics, however interesting they may be from the purely mathematical point of view.

On the other hand, the fact that atomic events are dominated by the elementary correlation law of unitary transformation, whereas ordinary games with dice and roulettes are not, may be taken as a sign that the microphysical quantum games deal with truly fundamental events pertaining to elementary particles rather than with complex heterogenous mechanisms.

28. *Identical particles and Nernst theorem*

We saw before that probability interference, usually regarded as a typical quantum feature, rests on a much broader basis, namely on the non-quantal postulate that a general law of interdependence between the various probabilities of transition exists at all. A similar situation prevails with respect to another feature, often considered as quantal, e.g. the 'unclassical' tendency of identical particles either crowding together in one and the same state (Bose-Einstein particles) or crowding out one another from the same state (Pauli-Fermi particles). The two modes of interaction are closely connected with the fact that the probability amplitude ψ as a function of the positions $q_1 q_2 \ldots$ of several identical particles is either *symmetric* or *antisymmetric* with respect to permutations of the particles. The symmetry rules (see below) which are responsible for the structure of atoms, molecules, and bodies in general, are said to involve 'some aspects of great philosophical content beyond the understanding of the physicist . . . who must be content to accept the implications without hope to penetrate the mystery that is implied' (Lennard-Jones[5]). Let us try to reduce the 'mystery' to common sense, in particular to the self-evident fact, which also may serve as a criterium of identity, that any observable quantity pertaining to a system of identical particles displays the same value whether particle a is at the place q_1 and b at q_2, or whether b is at q_1 and a at q_2, that is the probability postulate of symmetry

$$P(a_1 b_2) = P(b_1 a_2) .$$

From perturbation theory follows that the corresponding amplitudes ψ with respect to an exchange of places of a and b are

$$\text{either symmetric,} \quad \psi(a_1 b_2) = \psi(b_1 a_2)$$
$$\text{or antisymmetric,} \quad \psi(a_1 b_2) = -\psi(b_1 a_2) \ .$$

The consequences of either symmetry or antisymmetry on the constitution of matter are tremendous. The either-or, rather than both, follows from the thermodynamic necessity of dividing by $N!$, here by $2!$; refer to Section 21.

Even more simple is the proof that, in case of three (or more) identical particles a, b, c, ... in mutual interaction, ψ cannot be symmetric with respect to an exchange of a and b and at the same time anti-symmetric with respect to an exchange of a and c. Indeed, if it were so then one would have the sequence

$$\psi(a_1 b_2 c_3) = \psi(b_1 a_2 c_3) = -\psi(b_1 c_2 a_3) = -\psi(a_1 c_2 b_3)$$
$$= +\psi(c_1 a_2 b_3) = +\psi(c_1 b_2 a_3) = -\psi(a_1 b_2 c_3)$$

Since the result, $\psi(a_1 b_2 c_3) = -\psi(a_1 b_2 c_3)$ is self-contradictory, ψ can either be symmetric with respect to all three (or more) identical particles, or antisymmetric with respect of all of them. This means, however, that the class of all particles is divided in two subclasses: those which form symmetric ψ's (Bose particles) and those which form antisymmetric ψ's (Fermi particles). All this follows from the postulate that observable quantities ought to be symmetric with respect to any permutation of the identical particles, in short from the definition of identity as indiscernibility. It is far from self-evident, however, that the world contains uncounted billions of identical particles.

Symmetric and antisymmetric interaction implies that particles at the end of a long molecule, or at any distance whatsoever, 'know' what their identical partners are doing at the other end, so that they will be able to do the same as much as possible (Bose-Einstein) or not to do the same at all (Fermi-Dirac). As to the mutual communication over long distances, one must not forget that non-relativistic mechanics takes it for granted that interaction, even over planetary and interstellar distances, is instantaneous. A relativistic theory of interaction between several charged particles is still in the exploratory stage, in ordinary as well as in quantum mechanics. A relativistic quantum theory would also have to explain why particles of half-integral spin always prefer the antisymmetric mode of interaction. Pauli and Klein have recently made important forward steps toward a solution of this problem([6]).

The Nernst theorem, known also as the Third Law of thermodynamics, is connected with the symmetry or antisymmetry of the ψ-functions of N identical particles forming a body. Every state of such a body could be realized by $N!$ ψ-functions leading to an $N!$-fold 'degeneracy', if it were not for the symmetry requirement which reduces the number $N!$ to only *one* permitted ψ-function, either symmetric or antisymmetric.

The Nernst theorem must not be confused with the simple statement that one can never reach the absolute zero point $T = 0$. This empirical-looking statement becomes self-evident when one replaces the artificial absolute temperature scale T by the more natural scale $t = \log T$ where $t = -\infty$ stands for $T = 0$; and $t = -\infty$ can obviously never be reached. Chasing the absolute zero point is the Achilles chase in reverse. Achilles, of course, *reaches* the tortoise; only the trick of dividing a finite interval into an infinite number of steps makes it appear to be a surprising empirical fact that he actually succeeds. In contrast, the physicist of course *cannot reach* $t = -\infty$; only the trick of condensing the infinite t-interval into a finite T-interval makes it appear to be a surprising empirical fact that he actually will *not* succeed in reaching his goal. Considering the unattainability of $T = 0$ as a new law of thermodynamics is as misleading as regarding the fact that Achilles catches up with the tortoise as a new law of kinematics. Actually Nernst received the Nobel prize for a physical discovery about bodies approaching the (obviously unattainable) limit $T = 0$, namely, the experimental result that the entropies of various states of association and aggregation of particles forming a body converge toward a *common* entropy value differing by *finite* amounts from the entropies at higher temperatures. This is a very far-reaching empirical statement. Calling it the Third Law of Thermodynamics is not quite justified since it is a consequence of the fundamental symmetry principles as parts of the general ψ-metric.

The thermal properties of a mechanical system, in particular near the absolute zero point, depend essentially on the new statistics, either that of Bose-Einstein which accepts the symmetry of the ψ-functions with respect to the permutation of identical particles, or its opposite, the Fermi-Dirac statistics which only admits antisymmetric ψ-functions in agreement with the Pauli principle of excluding more than one particle per state. As an illustration take a 'gas' of three identical particles, a, b, c, distributed over three states belonging to the energies $E_1 E_2 E_3$. If $N_1 N_2 N_3$ are the occupation numbers of the three states, the total energy of the gas is

$$E = N_1E_1 + N_2E_2 + N_3E_3.$$

The entropy depends on the statistics by way of $S = k \cdot ln P$, where P is the number of possibilities to obtain a certain distribution with occupation numbers $N_1 N_2 N_3$. Permutations, of the identical particles are counted in full according to classical Maxwell-Boltzmann statistics; they are counted not as different in Bose-Einstein's statistics, that is only as *one* case ($P = 1$). In Fermi-Dirac statistics those distributions are admitted only for which none of the numbers N is more than unity. The three last rows of the following table give the values of P in case of the three statistics. The three first rows contain the occupation numbers:

N_1	3	0	0	2	1	0	0	1	2	1
N_2	0	3	0	1	2	1	2	0	0	1
N_3	0	0	3	0	0	2	1	2	1	1
M—B	1	1	1	3	3	3	3	3	3	6
B—E	1	1	1	1	1	1	1	1	1	1
F—D	0	0	0	0	0	0	0	0	0	1

Classical statistics counts the distribution $\begin{matrix} 2 \\ 1 \\ 0 \end{matrix}$ with $P = 3$ because it is realized by the following three distributions:

$$\left| \begin{array}{c} ab \\ c \\ \underline{} \end{array} \right| \begin{array}{c} ac \\ b \\ \underline{} \end{array} \left| \begin{array}{c} bc \\ a \\ \underline{} \end{array} \right|$$

which, according to Bose-Einstein are counted only as one, and by Fermi-Dirac as none. The exclusion principle applies to particles with half-integral spin, in particular to electrons.

Summary

From the two-way symmetry of the individual transition probabilities P follows that the P-tables must be unit sum squares. If one now postulates that there should be a general law of interdependence between the P-tables, symmetric and transitive, a law of order rather than chaos, then there is but one mathematical solution of the problem. It is known as *unitary transformation* with the help of intermediate complex quantities ψ arranged in matrices which satisfy the product theorem

$$(\psi_{AC}) = (\psi_{AB})(\psi_{BC}) \qquad \text{with } P = |\psi|^2.$$

(This triangular ψ-matrix connection is equivalent to a pentagonal connection between P-matrices, in analogy to plane geometry where ten lengths L connecting five points of a pentagon are always related so that nine L's uniquely determine the tenth length.)

Translated into physics, unitary transformation is identical with the *interference law of probabilities* of the quantum theory. This law, then, turns out to be a necessity under the postulate of order rather than chaos for the transition probabilities. The interference law does not have to be introduced as a sovereign quantum law. The result of this chapter is the second step in deriving the principal quantum features from a non-quantal basis, the first step being the two-way symmetry of the probabilities leading to the P-matrices as unit sum squares.

CHAPTER VII

ORIGIN OF THE QUANTUM
RULES

'Often we enter the unknown edifice of a new scientific discipline through a lesser gate that leads us into a side passage. It may take us a long while to find our way to the main portal and to view the whole structure in its proper perspective.'

SCHRÖDINGER

29. *Definition of dynamical conjugacy*

This chapter deals with the principal quantum rules, $E = h\nu$, $p = h/\lambda$, or rather their correct form, $\Delta E = h/T$ and $\Delta p = h/L$. Instead of accepting them as fundamental and irreducible, it will be shown that they are consequences of postulating invariance of mechanics with respect to shifts of zero points (Galileo and Lorentz invariance).

The quantum rules just mentioned and supplemented by the Bohr rule for the frequency of spectral lines, $\nu = (E-E')/h$, constituted the original Quantum Riddle. Today the quantum rules are derived from still more problematic general theorems, the Born qp-commutation rule, the Schrödinger p-operator rule, and (equivalent to the latter two) the theorem that the probability amplitude function $\psi(q, p)$ has the complex-imaginary form

$$\psi(q, p) = \text{const. } \exp(2i\pi qp/h) , \tag{1}$$

a wave function of wave length $\lambda = h/p$. Instead of giving us an explanation for all these strange rules, one expects us to accept them at face value 'because they work so well'. Could they perhaps be derived from a more fundamental basis?

The traditional answer to this question has been most unsatisfactory. For the last thirty-odd years the wave function (1) has been regarded as evidence of a dominant principle of nature according to which, using Newton's language, matter (and light) is 'endowed with an occult and specifick quality by which it acts and produces manifest effects', namely a dualistic interplay of wave and corpuscular qualities, attenuated by a fundamental complementarity. After so many years it is not even felt as problematic any more that matter sometimes displays particle, sometimes wave features. We are told that clicks in Geiger counters and Compton recoil effects can be explained only in corpuscular terms, whereas interference fringes in diffraction experiments can be accounted for only by wave action. To quote from an otherwise excellent textbook: 'Electrons, instead of having laws similar to classical laws, obey the laws of wave motion, ... and light is corpuscular in nature, at least when it interacts with matter.' It really looks like waves on Monday, Wednesday, and Friday, and like particles the rest of the week (Bragg). Only a few independent spirits, among them Einstein, Schrödinger, and de Broglie, have steadfastly refused to comply with this 'quantum mess' as final.

Instead of accepting duality as an 'occult and specifick quality' at face value, it is proposed in this study to 'derive it from two or three general principles of motion' (Newton's terms). One of them has been the non-quantal postulate that the probabilities of transition, arranged in unit magic squares, are connected by a self-consistent general law (Chapter VI); this law can have but one conceivable form, namely that of unitary transformation, identical with a wavelike *interference* of probabilities. The other 'occult and specifick' quantum feature, the periodic relation between *conjugate* dynamical observables q and p, will be deduced in this chapter.

Conjugacy between quantities q = co-ordinates and p = momenta has been defined differently in classical and in quantum mechanics.

(a) *Classical mechanics* defines q and p as conjugate pertaining to a certain mechanical system when the equations of motion have the canonical form

$$dq/dt = \delta H/\delta p \text{ and } dp/dt = -\delta H/\delta q.$$

This definition refers to energy and time, which themselves represent a conjugate pair. The classical definition thus is circular.

(b) *Quantum mechanics* usually defines q and p as conjugate coordinates and momenta when they satisfy the Born commutation rule, $qp - pq = h/2i\pi$, or the Schrödinger operator rule, $p = (h/2i\pi)d/dq$, or when they give rise to the wave function (1). All three conditions introduce the quantum relation between q and p by decree, thus putting quantum dynamics on an *ad hoc* basis.

Our task is that of deriving quantum dynamics as a consequence of simple non-quantal postulates and in particular obtaining the periodic relation between q and p without introducing special quantum features, such as (b).

It is characteristic of mechanics that, whenever it deals with a co-ordinate value q, this value is meant with respect to a certain zero point. Hence, when one has to do with two values q and q', only their difference is of physical significance, and the arbitrary zero point cancels out in $q - q'$. The same holds for the momentum p, as well as for energy E and time t. In short, mechanics is Galileo invariant. Transferring this invariance from classical to quantum mechanics, we now introduce the postulate:

(c) Any observable $T(q)$ has matrix elements (= transition values) $T_{pp'}$ which depend on the difference $p - p'$ only. Similarly, any observable $S(p)$ has matrix elements $S_{qq'}$ depending on $q - q'$ only. Analogous results hold for E and t.

The postulate (c) may be used as a definition of the conjugacy of p and q, and also of E and t, in place of the customary quantum definition (b) above. In contrast to (b), the postulate (c) does not contain any quantum feature of pq-periodicity. On the contrary, the main topic of the present chapter is to show that, from the non-quantal postulate (c) of invariance of mechanics with respect to shifts of zero points one can *derive* that the probability amplitude function $\psi(p, q)$ must have the wavelike periodic form quoted in (1) above, as the third step in deducing quantum mechanics from a non-quantal basis.*

* Introducing a relativistic denominator as in (6) p. 165 implies a change of the foundations.

30. *Origin of the wave function*

The proof that $\psi(q, p)$ must be of the complex periodic form (1) rests on answering the third of the following questions:

1. Which function $\phi(p)$ satisfies $\phi(p) + \phi(p') = \phi(p \cdot p')$?
Answer: the logarithmic function, since $lg\, p + lg\, p' = lg(p \cdot p')$.

2. Which function $\chi(p)$ satisfies $\chi(p) \cdot \chi(p') = \chi(p+p')$?
Answer: the exponential function, since $e^p \cdot e^p = e^{p'+p'}$.

3. Which function $\psi(p)$ satisfies $\psi(p)\psi^*(p') = \psi(p-p')$ where the asterisk may or may not signify the complex conjugate?
Answer: the complex exponential function, e^{ip}, since $e^{ip}\, e^{-ip'} = e^{i(p-p')}$, so that the asterisk means complex.

It is shown in mathematical detail in Appendix 3 that, on the grounds of postulate (c), the probability amplitude function $\psi(p, q)$ must necessarily have the periodic complex form

$$\psi(p, q) = \text{const} \cdot \exp(2i\pi pq/\text{const}) = \psi^*(q, p), \qquad (2)$$

with the constant in the exponent denoted by the letter h. It is a periodic function in q-space of wave length $\lambda = h/p$. For the conjugate pair E and t one obtains the analogous result

$$\psi(E, t) = \text{const} \cdot \exp(2i\pi Et/h) = \psi^*(t, E) \qquad (3)$$

periodic in time with period $\tau = h/E$.

(2) and (3) lead further to the quantum rules of momentum and energy exchange (see below). The basic quantum theorems (2) and (3) together with the quantum rules can thus be derived from general non-quantal postulates of symmetry and invariance. The authoritative ([3]) statement: 'We know today that the quantum rules are consequences of a basic fact of all atomic events: the dualism of the wave picture and the particle picture' ought to be changed to read: 'We know today that the quantum rules are consequences of a basic fact of all events, not only atomic ones, namely (1) the two-way symmetry of the probabilities, (2) the postulate of a general interdependence of the probabilities, and (3) the

invariance of mechanics with respect to shifts of zero points.

It is nothing new that from the periodic function (1) follows the Born commutation rule

$$pq - qp = ih/2\pi \qquad (4)$$

of operator mechanics, and the special operator

$$p = (ih/2\pi) \, \partial/\partial q \qquad (5)$$

of wave mechanics. It must be kept in mind, however, that the enigmatic-looking commutation rule, and operator calculus in general, is but a shorthand way of writing connections between data which could as well be written in a less symbolic and abstract form.

For later use notice that the general tensor transformation formula

$$T_{pp'} = \sum_q \sum_{q'} \psi_{pq} \, T_{qq'} \, \psi_{q'p'}$$

reduces for a function $T(q) = T_{qq'} \, \delta_{qq'}$, and because of (2), to

$$T(p, p') = \int T(q) \exp [2i\pi(p-p')q/h)] \, dq \ . \qquad (6)$$

The question has been asked, why all particles have the same action constant h. We only can answer that, if different particles were dominated by quantum rules with different constants h, there could not be any (quantal) energy and momentum exchange between them. There simply is only one world rather than several interwoven independent worlds. Another question is: Why has h so small a value as $6 \cdot 54 \ 10^{-22}$ in centimeters, grams, and seconds? Answer: This safeguards the approximate validity of deterministic mechanics in macroscopic dimensions without which an organic structure including theoretical physicists could not exist.

The following remarks may further illustrate the nonquantal basis of quantum mechanics. The noncommittal postulate that there be a general law of interdependence between various probability tables leads to the specific feature of probability interference, best expressed in terms of probability amplitudes ψ. Still, the interference law is of little help for the prediction or calculation of experimental results: even when two P-matrices (P_{AB}) and (P_{BC}) are ascertained experimentally, this does not suffice to calculate the matrix (P_{AC}) because the P-measurements in the two first matrices leave the ψ-phases undetermined. This situation, encountered in the preceding chapter, is radically improved by the present one. The one additional non-quantal postulate that dynamics be invariant with respect to shifts of zero points in q- and p-space leads to the determination of *one* ψ-table or ψ-function, the wave function $\psi(q, p)$, and similarly to $\psi(E, t)$. After the one basic function $\psi(q, t; p, E) = \exp[2i\pi(pq \pm Et)/h]$ is thus known from the theory, all other ψ-functions, such as $\psi(A; q, t)$ and $\psi(B; q, t)$ and then $\psi(A, B)$ together with the eigenvalues of the observables A and B can be calculated when A and B are defined as functions of q and p, and/or E and t, by means of the ψ-interference theorem which, by way of the operators p and E (see above) lead to the Schrödinger equation. We do not have to go into details here since these developments are treated in the textbooks. However, the books usually take the pq-periodicity (and the pq-exchange rule and p-operator rule which follow from it) as an axiom which cannot be further reduced, whereas we *derive* the quantum rules from general non-quantal postulates of symmetry and invariance. That is, the textbooks begin where our developments end.

The only other attempt, as far as I know, of deriving the quantum rules from a non-quantal basis is due to F. Bopp[2]. He introduces sixteen formal postulates, some of a highly mathematical character, whereas our aim is to introduce only simple and plausible ground postulates of an immediately obvious physical significance.

31. *Periodicity in one dimension*

The quantum rules, $\Delta E = h/\tau$, $\Delta p_\phi = h/2\pi$, and $\Delta p = h/L$ were originally regarded as sovereign principles which can

not be further explained. Actually, they are but special consequences of the general probability metric of unitary transformation (probability interference) together with the invariance postulates of mechanics which lead to the periodicity
of $\psi(p, q)$ and to analogous functions $\psi(E, t)$ and $\psi(p_\phi, \phi)$.
Let us derive a quantum rule for the momentum activity of a
body. We consider first the case of a 'one-dimensional crystal'
periodic in the one linear direction q so that those observables
$T(q)$ which depend on q alone have the same values at q as at
$q + l$ and $q + 2l$, and so forth. $T(q)$ may then be expanded
as a Fourier series with linear periodicities l/n:

$$T(q) = \sum_n T_n \cos (2\pi nq/l + \alpha_n) \qquad (7)$$
$$= \sum_n \tfrac{1}{2}T_n \exp (\pm 2\pi inq/l \pm i\alpha_n)$$

where the $\pm$ indicates a sum of two terms, one with $+$ and
the other with $-$ sign. In order to investigate the momentum
activity of this periodic body we substitute (7) into (6) and
obtain

$$T(p, p') = \sum_n \tfrac{1}{2}T_n \int \exp \left[2i\pi \left(\frac{p-p'}{h} \pm \frac{n}{l} \right) q \pm i\alpha_n \right] dq \quad (8)$$

Since the integrand is periodic in q, the integral consists of
positive and negative contributions which cancel one another
unless the factor of q in the exponent vanishes. $T(p, p')$ has
non-vanishing values only for differences

$$\Delta p = p - p' = \pm nh/l. \qquad (9)$$

For values of Δp not satisfying this condition the transition
values $T(p, p')$ vanish. The physical meaning of this result
is this: Our one-dimensional crystal is, from the dynamical
point of view, a body which can change its momentum p in
the q-direction only in quantized amounts, $\Delta p = \pm nh/L$,
that is in multiples of the basic periodicity L along the q-axis.

In Chapter I we have seen that the quantum rule for the
exchange of linear momentum leads to a complete explanation

of the diffraction of electrons in selected directions from a crystal by means of Duane's equation of momentum balance

$$2p \sin \theta = nh/l \quad \text{(Duane)} \tag{10}$$

which is equivalent to the wave interference relation

$$2l \cdot \sin \theta = n\lambda \quad \text{(Bragg)} \tag{11}$$

by virtue of a translation of the physical particle momentum p into a wave length $\lambda = h/p$ of hypothetical matter waves—only that the quantum mechanics of the third quantum rule disposes of the need for translation or double manifestation once and for all, although it may remain as a convenient ideological bridge to get quick results.

32. *Periodicity in three dimensions*

A three-dimensional crystal is composed of basic cells characterized by three ground vectors a_1 a_2 a_3 filled with matter according to a density pattern repeated in every cell. The direction of a set of parallel lattice planes is defined by a plane laid through the endpoints of three vectors n_1a_1, n_2a_2, n_3a_3 with integers n_1 n_2 n_3 which also determine the distance between subsequent planes. The Bragg wave condition (11) again determines the angles of selected reflection according to wave interference theory.

The same selected angles can be obtained also from a construction in the 'reciprocal lattice', the basic cell of which is formed by three ground vectors b_1 b_2 b_3 which are perpendicular to the walls of the former space lattice, and whose magnitudes are reciprocal to the perpendicular distances d_1 d_2 d_3 between the walls of the space lattice cell

$$b_1 = 1/d_1, \quad b_2 = 1/d_2, \quad b_3 = 1/d_3$$

A point of the reciprocal lattice is reached from its zero-point by a vector

$$R = k_1b_1 + k_2b_2 + k_3b_3$$

H

with three integral numbers k_1 k_2 k_3. The reciprocal lattice is presented by Fig. 6 with O as zero-point.

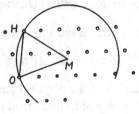

Fig. 6

The following construction for the selected reflection of incident waves λ has been given by P. P. Ewald([3]). Draw a vector MO of length $1/\lambda$ in the direction of the incident waves ending in a point O of the reciprocal lattice. Then draw a sphere around M with radius OM. If this sphere runs through, or very close to, another point of the reciprocal lattice, such as the point H, draw the vector MH; it will represent a reflected direction, when MO is the incident direction. Geometrical considerations show that this construction is equivalent to that of Bragg.

Ewald's construction may be interpreted as a construction in momentum space when multiplying all lengths in the diagram of Fig. 6 by the factor h. The vector MO then represents the momentum of the incident particle of magnitude h/λ, whereas MH is its momentum vector after reacting with the crystal. The magnitude of the particle momentum remains unchanged during the reaction with the crystal; only its direction is changed according to the momentum vector equation $MO + OH = MH$. Hence OH represents the momentum vector gained by the particle. A crystal can thus be described as a mechanical system which is capable of giving out (or receiving) only such directed momenta as are represented by connections between any two points in the reciprocal lattice, multiplied by h. The two descriptions of the crystal, as a periodic arrangement of matter in space, and as a mechanical system giving out and receiving quantized momentum vectors, are equivalent. However, according to the unitary particle theory of matter, only the second description is a realistic repoit on the diffraction of *matter*; the wave description deals with probability amplitudes of matter particles, not with real waves. We cannot discuss photons *versus* light waves since this would lead into details of the quantum theory of fields.

Ewald's construction and its mechanical interpretation after multiplication by the factor *h* is identical with Duane's third quantum rule of mechanics. The fact that this rule has never been recognized in its significance for a *unitary* particle *interpretation* of wavelike phenomena such as diffraction through two slits via periodic space components of the diffractor, is one of the strangest items in the history of the quantum theory.

33. *Periodicity in time*

It is known from ordinary mechanics that energy *E* and time *t* play the part of conjugate observables similar to *p* and *q*. One only has to substitute *E* for *p* and *t* for *q* in the above developments to arrive at new consequences of our non-quanta ground postulates. The first conclusion is that the amplitude function $\psi(E, t)$ must have the form

$$\psi(E, t) = \text{const} \cdot \exp(2i\pi Et/h) \tag{12}$$

Suppose, further, that those observables of a certain object which are functions of *t* have the simple harmonic form

$$T(t) = T_0 \cdot \cos(2\pi\nu t + \alpha)$$

The object would then be called a 'harmonic oscillator' of frequency ν. Another object (atom) may have observables $T(t)$ of the more involved form

$$T(t) = \sum_n T_n \cos(2\pi\nu_n t + \alpha_n) \tag{13}$$

Substitution into the integral:

$$T(E, E') = \int \psi(E, t) \, T(t) \, \psi(t, E') \, dt$$

then yields the result, similar to (9), that $T(E, E')$ has non-vanishing values only for differences

$$\Delta E = E - E' = \pm h\nu_n \qquad (14)$$

This is Bohr's frequency condition. Thus, when observing the emission or absorption spectrum ν_1 ν_2 ν_3 ... of our object (atom), we can infer from it a set of energy levels characteristic of the object, which yield the observed frequencies according to the Bohr frequency condition. (1) and (12) lead to the quantum prescriptions of Born and Schrödinger, to Heisenberg's uncertainty relation, to wavelike appearances, and all the rest. If these matters were once beyond our understanding, if they were 'riddles', then they have now been solved by reduction to almost obvious non-quantal postulates.

A serious paradox connected with Planck's quantum rule has been seen in the contradiction between *sudden* energy changes $\Delta E = h\nu$ of an oscillator and the *long-lasting* coherent light vibration initiated by it. It must be emphasized here that Planck's relation does not imply a sudden energy change of the oscillator. It rather describes the result of an interaction between the oscillator ν and a field component ν spread out in space. There is a large *probability* that the reaction leads to energy changes $\pm h\nu$ of the two participants observed after their separation. Refer to the footnote on p. 16.

Summary

The quantum periodicity rules connecting coordinates and momenta, as well as energy and time are derived from the following assumptions:

(a) Let the two quantities p and q be variable between $-\infty$ and $+\infty$. Let the components $T_{pp'}$ of every tensor $T(q)$ in the space of unitary transformation depend only on the difference $p - p'$, and the components $S_{qq'}$ of every tensor $S(p)$ depend on $q - q'$ only. This leads to the conclusion that the unit tensor ψ must have mixed components $\psi_{qp} = \psi(q, p)$ = const $\cdot$ exp$(2i\pi qp/$const$)$. The same for E and t.

(b) The quantities q and p are identified as linear coordinates and conjugate linear momenta, and the tensors S and T as observables. The constant in the exponential function ψ is known as Planck's action constant, so that $\psi(q, p)$ is a 'wave function' of wave length $\lambda = h/p$.

The wavelike quantum periodicity is thus explained as a consequence of familiar features of q and p in classical mechanics, namely *invariance* with respect to Galileo transformation, combined with the general theorem of unitary transformation which alone produces a general connection law between the unit-sum-square quality of the P-tables, which in its turn rests on the two-way *symmetry* of the transition probabilities.

CHAPTER VIII

QUANTUM FACT AND FICTION

'The history of every science is littered with the bones of dogma proposed by some exceptional scientist and, for a time, widely and enthusiastically acclaimed by the less enterprising members of the discipline.'

R. LAPIERE

34. Remarks on Method

There is consensus among the experts concerning the actual solving of atomic problems by the mathematical methods of quantum mechanics. Differences have arisen only with respect to interpretation and ideology. But ideology is a serious matter for the adept and the interested public at large. One wishes to have a concrete understanding why one must resort to the puzzling mathematical formalism of noncommutative algebra and complex imaginary functions which has been so successful in coordinating diverse phenomena in the microphysical realm. Having dealt in Chapter I with the dualistic ideology and its repudiation by the unitary quantum mechanics of particles, we are now going to wage a cold war against many other notions developed during the creative stage of the 1920's and early 1930's, notions which have shown a remarkable staying power in spite of their problematic character and are now sanctioned by authority and tradition to such a degree that critique is shrugged off as thinking unthinkable thoughts. Clarity has been particularly obscured by a linguistic approach to problems of physical interpretation so that Niels Bohr is praised because he alone 'knew that the real problem . . . of eliminating contradictions . . . was to refine the *language* of physics' (refer to Section 44). At the other extreme, there has been an inclination to see the essence of atomic science in the abstract mathematical symbols and their manipulation, as expressed by Eddington: 'We have learned that the exploration of the external world by the methods of physical science leads not to a concrete reality but to a shadow world of symbols . . . as one of the significant recent advances so that *homo sapiens* merely finds his own footprint in the sand'. How from this shadow world of symbols Man could ever have developed his technical domination of natural forces remains unexplained, even when we learn from another source that only *modern* science has become a world of symbols: 'The first step in science's retreat from mechanical explanation to mathematical abstraction was taken in 1900 when Max Planck . . .', etc. As though $E = \frac{1}{2}mv^2$ were mechanical explanation but $E = h\nu$ were retreat to mathematical abstraction. Still, whether retreat or advance, a shadow world of symbols would signify a complete reversal of natural science, the object of which was, and still is: to suppose that a real world exists without human advice and consent, then to search for general regularities which may help us to manipulate things and events. Philosophers have raised the question of how we can arrive at any knowledge of an 'external world' without mental ingredients. If they have leaned toward idealism or positivism, or other isms, this goes beyond physics. It seems idle, however, to appeal within physics to a philosophical doctrine in order to find an easy way out of a purely scientific problem such as whether matter really consists of particles or of waves. It has been asked: 'What do you mean by matter particles being real and matter waves being mere appearances?' To the scientist a 'real thing' is characterized by certain constantly recurrent qualities. An electron with its discrete and condensed charge, mass, and spin is a real thing. But the probability curve or ψ-function describing betting odds for future events is not a real thing, even when the curve looks wavelike. Using the good Doctor Johnson's language: You can kick a stone, and you also can kick an electron, and even a water wave and an electromagnetic wave, and be hurt by them, proving their 'reality'. But you cannot kick, or be hurt by, a wavelike curve representing probabilities of events. Future historians of science will have a hard time to understand how present-day quantum physicists, under the incessant barrage of their leaders' dogmatic pseudo-empiricism, have become blind to this simple distinction between the real and the apparent within their own science.

If one accepts the statistical interpretation of Duane and Born, then most of the philosophical profundities to justify the illusion of duality are superfluous. One then can return from subjective pictures and shadow symbolism to plain physics again and speak unashamedly of electrons as real things without philosophical trimmings. On the other hand, if Bohr and Heisenberg are severely criticized in this chapter for their subjectivistic approach, their language refinement, and their occasional deviation from logic and consistency, it will hardly be necessary to add that my admiration for their achievements in theoretical physics is as great as ever. But I must confess that, during many years of trying to imbue my students with the Copenhagen Spirit, I felt more and more like the Devil's Advocate, suffering from an ever-growing intellectual distress. Also, if the considerations of this last chapter will resound again and again, like a *basso continuo*, with Duane's third quantum rule, the indulgence of the reader may be asked on the grounds that this most fundamental quantum law has been consistently disregarded throughout the last three decades in dozens of symposiums, scores of technical and popular books, hundreds of learned articles, and thousands of classrooms, with catastrophic results for the interpretation of atomic theory, ending up in 'a quagmire of evasion', using William James's famous phrase. To quote just one short example, the prominent German quantum philosopher, von Weizsäcker[1], enlightens his readers as follows: 'We know today that [the quantum laws $E = h\nu$, etc.] are the consequence of a basic fact of all atomic events: the dualism of the wave picture and the particle picture.' Even if it were true that the quantum laws could be shown to be consequences of duality, it would only be a derivation of physical laws from an unsupportable antinomy which, as we know today, is *not* a basic fact. Nor is the duality antinomy a consequence of the unitary quantum laws.

35. *Duality and doublethink*

There have been various attempts at constructing a unitary theory of matter, the most notable one being Schrödinger's unitary wave theory resting on the idea that what appears to be a particle is in reality the high crest of a wave. Indeed, according to the wave equation, such a wave crest moves on the same path and with the same velocity as a particle would move according to the equations of mechanics. The drawback is that a high wave crest flattens down after a short time, thereby losing its particle-like condensation altogether, so that this idea had to be abandoned. The counterproposal, a unitary mechanical particle theory, prepared by Duane's mechanical explanation of diffraction, was reaffirmed by Born in the general statistical interpretation of Schrödinger's wave function. There is hardly a physicist today who would oppose the Born interpretation. Strangely enough, in their idle hours many theorists still pay lip-service to the archaic idea that particles and waves of matter are pictures on an equal level, rather than conceding that particles, according to modern quantum mechanics, are the real constituents of matter, and the waves are mere appearances produced by the statistical co-operation of many particles.

Resignation to an inscrutable duality of manifestations may have been reassuring to an earlier generation in despair over an apparent antinomy. But later the temporary resignation turned into a new *credo*, in particular promoted by Heisenberg's famous Chicago lectures[2] of 1930, illustrated by two photographic records, the one showing corpuscular tracks in a cloud chamber, the second displaying an electronic diffraction pattern with maxima and minima of intensity. But it cannot be stressed too much that those maxima are produced by individual particle impacts in statistical distribution, so that *Heisenberg's second exhibit proves the opposite of what it was supposed to prove*: it confirms Born's statistical particle interpretation of wavelike phenomena of matter. There is no need for duality any more. '*Je n'ai pas besoin de cette hypothèse*' said Laplace on another occasion. Clinging to dualism violates one of the principal rules of orderly thinking: Do not indulge in false opposites. Do not construe an antithesis between a thing, a bundle of qualities, as against *one* of these qualities. Do not contrast a snake as a thing with its occasional wavy shape in order to defend a duality of snake substance. Quantum theorists, however, fall short of the standards of logic when they proclaim an opposition between particles as the substance of matter on the one hand, and the occasional wavelike statistical disposition of the same particles on the other. If the experts cling to their accustomed language and use alternate methods for heuristic purposes, there is no harm. But the constant harping on an illogical antinomy does not promote clarity, as a glance at the literature reveals. Thus Sir James Jeans in his Presidential Address to the Royal Society proclaimed in 1934([3]): 'The particle picture is a materialistic picture which caters for those who wish to see their universe mapped out as matter existing in space and time. The wave picture is a determinist picture which caters for those who ask the question: What is going to happen next? The wave picture which observation confirms in every known experiment exhibits a complete determinism . . . but it is one of waves and so, in the last resort, of knowledge. As to energy conservation, waves of knowledge are not likely to own allegiance to this law.' One will not blame Jeans for his distinguished contribution to the general confusion prevalent thirty years ago. But in 1957, when the dust should have settled, we read in another presidential address([4]): 'You can only know fully the details of half the things in the world, . . . and this means that you can never devise an experiment which will say: this is a particle and not a wave, or vice versa.'

Actually, to those familiar with the third quantum rule, every experiment, including that of matter diffraction says: This i a particle and nothing else. And the theory of quantum mechanics says the same.

Looking through bifocal glasses, von Weizsäcker[1], puts it this way: 'In what respect does quantum mechanics differ from classical physics? It has discovered that the same physical object, for example an electron, appears in two seemingly different forms . . . wave or particle . . . cloud-chamber photographs versus interference. . . . Now what meaning can there be in the statement that an electron is both particle and field?' The logical answer to this query is: 'no meaning at all' since an electron *is* a particle, but the statistical behavior of many electrons is not what an electron is. Later the same author maintains: 'The same atom behaves in some experiments like a spatially contracted particle, in others like filling the whole space.' If this were the voice of science, it would be rather alarming. Actually, von Weizsäcker only repeats the catechism of a past generation and ignores the scientific lesson of Duane and Born. Indeed, according to *modern* physics, atoms and electrons are particles not only by virtue of their condensed charges and masses. Even in their statistical display they behave exactly as particles ought to behave. For, although interference looks wavelike, it is a necessity under the probabilistic quantum mechanics of particles (Chapter VII). When von Weizsäcker finally declares: 'Planck's law $[E = h\nu]$ is the expression of the relation between two pictures,' he is, like Rip van Winkle, at least two generations behind the times. For already in 1900 Max Planck taught us that $E = h\nu$ does *not* express a relation between two pictures but rather describes a fundamental *physical* fact, namely that a system of time period T changes its energy only in amounts $\Delta E = h/T$. A few years later Sommerfeld and Wilson found the corresponding physical fact that a system of angular periodicity 2π changes its angular momentum only in amounts $\Delta p_\phi = h/2\pi$. (By the way, one would like to know whether this rule, too, expresses a relation between two pictures, and what these pictures are.) Finally, in 1923 Duane added the third rule that a body with linear periodicity L in space changes its linear momentum in amounts $\Delta p = h/L$. If the latter rule was lost in a medley of subjective interpretations, one would expect at least that the first two quantum rules are recognized as objective laws of physics, rather than as relations between two pictures.

It is true that originally there was no convincing explanation of the quantum rules; they constituted a Quantum Riddle. But von Weizsäcker goes much further and tells us: 'In quantum theory these relations play the role of axioms which are suggested by experience, but can no longer be deduced from other laws.' So everything is fine; *do not disturb*. Actually, the quantum rules of energy and momentum exchange can very well be deduced from other laws, as shown in the preceding chapters. If von Weizsäcker has been singled out in our criticism above, it is only because he, as a pupil of Heisenberg and as a recent adept of the philosophy of science, supposedly speaks with authority.

Strangely enough, even Max Born to whom we owe the statistical interpretation of the Schrödinger ψ-function as a probability amplitude for *particles*, makes belated concessions to dualism. In an article on 'Physical Reality' in 1959[5] he points out that the popular concept of a real thing requires the presence of certain conservative traits, conceived as 'Gestalt' in ordinary life, and as invariant quantities in physics. 'Invariants are the concepts of which science speaks in the same way as ordinary language speaks of "things" and which it provides with names as though they were ordinary things.' Electrons are real things because they display invariant rest mass, spin, and charge, even though they *supposedly* do not have simultaneous positions and velocities (see Section 37, however). In the same way, water waves are real things. But then Born continues: 'Why then should we withhold the epithet "real" even if the waves represent in quantum theory only a distribution of probability?' My answer is: For the same reason that sick people are real things, whereas the wavelike curve which symbolizes the probability for a person to be sick during a fluctuating epidemic is not a real thing. Regarding a wavy disease curve as a thing as real as a sick person, and speaking of the 'symmetry' of disease curves and sick persons is nonsense. In the quantum domain, however, the declaration 'why should we withhold the epithet "real" . . .', etc., promotes a comfortable *chiaroscuro*, giving the impression that an elusive but profound ambiguity is at stake which must be protected from profane criticism at all costs. Refer to the incisive critique by W. Yourgrau ([25]).

Niels Bohr speaks of 'the essential element of *ambiguity* involved in ascribing conventional physical attributes to atomic objects, as is at once evident in the *dilemma* regarding the corpuscular and wave properties of electrons. . . .'([6]) Ambiguity? Dilemma? Both have been removed by the third quantum rule! In 'Words and Things' Roger Brown writes: 'There is a familiar inclination on the part of those who possess unusual and arduously acquired experience to exaggerate its remoteness from anything the rest of us know.' These words come to mind when Bohr comments on the ψ-function and on quantum mechanics in general: 'These symbols, as is indicated already by the use of imaginary numbers, are not susceptible to pictorial interpretation.' This is yielding to sixteenth-century word magic. For why should 'imaginary' numbers be less pictorial than 'irrational' or 'transcendental' numbers? Imaginary numbers are but shorthand symbols describing relations in a plane, pictorial or not. On the other hand, it would indeed be too pictorial to visualize a travelling electron as a tiny ball with wavy wings attached, as in a certain sports emblem. Yet this very picture is suggested by the often heard phrase that an electronic particle, baptized 'wavicle', is *associated* with a wave phase and amplitude, symbolized by a complex ψ-function. This rests on a misunderstanding of 'wave mechanics'. An electron is there as a physical entity, and it may be found in a certain physical state A. This state

(geometrically: a vector in the space of unitary transformation) can be represented by an infinite variety of ψ-function,
signifying probability relations between the one state A and
an infinite number of mutually orthogonal sets of other states,
each such set being described by one ψ-function. Hence one
state A can be represented by an infinite number of different
ψ-functions, each of them containing in its turn an infinite
number of phases and amplitudes. Hence it is entirely misleading to imagine an electron in a certain state A as being
'associated' with any one set of phases and amplitudes, let
alone with one single phase and amplitude. The latter idea
goes back to the picture of waves carrying along electrons, as
the wind carries along dust particles. But after three decades
this primitive picture still figures in discussions about 'Observation and Interpretation'[4] after having been revived by
de Broglie, Bohm, and Vigier.

It may seem pedantic to criticize such antiquated notions
which still may have heuristic value. The trouble is that the
student becomes imbued already as a schoolboy with the
alleged dual nature of matter as the great novelty of modern
physics. Thereafter he has to pass through a long period of
bewilderment until he finally learns to grope his way more or
less instinctively through what many denote as 'the quantum
mess', without receiving much help from his likewise infected
teacher.

36. Neo-dualism, photons

After Duane's particle mechanics of diffraction and Born's
general statistical interpretation had once been established, it
should have become clear that *all* phenomena of matter can be
accounted for by the unitary particle mechanics. Yet dualism
is still defended by the argument that *all* phenomena can also
be accounted for by the theory of a fluid matter continuum
known as *Second Quantization* (Klein, Jordan, and Wigner,
1928), obtained by a coordinate transformation from the customary First Quantization of particle mechanics. Since the
two quantizations are but different mathematical forms de-

scribing the same observations, why should we regard the usual particle theory as representing the 'true' situation any better than the equivalent theory of a quantized continuous fluid? Does this not make the world safe for duality again? I do not think so. The scientific value of a theory is measured not only by its power to account for observed data but also by criteria of simplicity, freedom from *ad hoc* assumptions, and reducibility to more general postulates. For example, the motion of the planets has been described by Ptolemy, Tycho, and Copernicus by three models differing only as to their reference points in space. Yet the Copernican system is far superior, and thus, scientifically speaking, represents the true situation, not only by virtue of its formal simplicity but even more by its reducibility to the general mechanics of Newton —whereas circles upon circles can at best be justified on metaphysical grounds that celestial bodies, aiming at perfection, must properly move on the most perfect of curves. Speaking of planetary triality does not make sense.

A similar situation prevails with respect to the First and Second Quantization. The former regards an atom as consisting of N discrete electrons surrounding a nucleus and controlled by quantum rules of mechanics which can, as seen in the previous chapters, be reduced to simple and general postulates. The Second Quantization, however, describes an atom as a continuous electric fluid dominated by an utterly complicated non-linear differential equation. Imagine for a moment that it had been discovered first that the qualities of the ninety-two natural atoms could be understood in terms of ninety-two different electric fluids, controlled by an intricate hydrodynamic equation, whereupon someone discovered that this cumbersome formalism can be rewritten in a simple way so as to depict the model of N electronic particles held together by an inverse square law of attraction and repulsion. Then Mr X would certainly be hailed as the Copernicus, Galileo, and Newton combined of the microcosm—rather than as the inventor of another equivalent mathematical schema. This consideration alone condemns the notion that the (otherwise most valuable) K-J-W transformation has saved duality. From any sound scientific point of view, an Argon atom *really and truly* consists of eighteen electrons surrounding a center of charge number 18. Saying that it can 'as well' be regarded as an intricate electric fluid is not a matter of taste; it is the ultimate betrayal of Galilean science, using a phrase of K. Popper. Besides, it is an error of misplaced emphasis, like classifying Man as a featherless biped, because the K-J-W formalism, does not contain waves λ and ν corresponding to an electronic momentum p and energy E in the Rutherford model. Denoting as 'dualism' the juxtaposition of K-J-W to the quantum mechanics of electrons is using a once respected trade name for an entirely different Ersatz Article, sold without proper notification of the customer, that is the listener to inaugural addresses, the visitor of science pavilions at world fairs, and the reader of books on 'The Revolution of Thought in Modern Physics'. Therefore, when Heisenberg maintains([7]): 'The symmetry between waves and particles . . . has to be regarded as an essential feature of quantum theory since Bohr's work of 1927 and since the investigation of Klein, Jordan, and Wigner', this ought to be taken with at least two grains of salt because Bohr's work of 1927 was repudiated by Duane's work of 1923, and K-J-W's continuum does not produce even a semblance of symmetry between waves and particles since it does not contain waves.

Of course, nobody will deny that the original candid dualism has been a most fertile heuristic idea. Nor am I blind to the fact that the K-J-W- transformation as such has great merit for the quantization of a continuum, for example, of an electromagnetic field in three-dimensional space. But within the theory of matter it merely offers a transition from the simple mechanics of the Rutherford model of the atom to a complicated mathematical form that could as well be augmented by many other mathematical representations, supporting a doctrine of plurality rather than duality. Still, it is the method of science to accept one exceptionally simple form of the theory as the 'true' one. Classical mechanics, too, can be written in various equivalent forms, as differential equations of motion, or as theorems of economy (variation principles) without claiming duality or plurality. Taking the K-J-W transformation as another argument for dualism is but a subterfuge to protect the fallacy of the original duality based on ignoring the third quantum rule of mechanics.

Turning now from the theory of matter to that of light, one may ask again: which is real and which is mere appearance, light waves or photonic particles? The reply here is just opposite to that of matter. Photonic particles, once introduced by Einstein as a heuristic hypothesis, are abandoned in the modern theory of fields. The substance of light is a continuous field which carries energy, hence mass. It is divided into harmonic components or field oscillators, each of them characterized by a generalized coordinate q and momentum p, subjected to the same quantum periodicity rules as the q's and p's of material oscillators, although each field oscillator is spread out continuously in space, rather than being condensed in a small section of space like a particle. W. Heitler, in his standard work *The Quantum Theory of Radiation*[8], often speaks of light quanta or photons. But he significantly adds: 'Light quanta appear in the theory only as quantum numbers attached to the radiation oscillators.' But quantum numbers do not dash around like electrons. The real constituents of matter are discrete particles, occasionally giving the appearance of wave action. And the real constituent of light is a continuous electromagnetic field, sometimes giving the appearance of photonic particles. This is the present view which may not be eternal, of course.

Theorists, however, with their peculiar quantum logic (*honi soit qui mal y pense*) pretend to have forgotten the difference between what things are and what they seem to be. Having attended a compulsory course in Philosophy Hall, they have gathered that 'things' are but mental constructs fashioned from bundles of sensations, that there is no physical reality in any objective sense, *an sich*. Maybe so. Yet, within physical science as well as in ordinary life, the distinction between reality and appearance is vital. A cloud may appear as a continuous ball of white stuff, though in reality consisting of many small droplets. Similarly, electric matter may sometimes give the appearance of a continuous medium supporting waves; yet in reality it consists of discrete particles, even in case of diffraction. Electromagnetic fields with their oscillating components are real, and they interact physically with other things. But the ψ-functions which, in the quantum theory of radiation, describe probabilities of field oscillator intensities are not real (kickable) things. Appealing to arguments in the domain of idealistic philosophy, just in order to avoid a clear choice between particles and waves, does not live up to the standards of scientific method with its First Commandment, Occam's rule: 'Entities must not be multiplied beyond necessity.' There is a big difference between (a) the fact that statistical distributions and particle energies can be calculated by the wavelike equation of Schrödinger—a tremendous scientific achievement, and (b) the assertion that matter displays symmetry of particle and wave manifestations, together with philosophical implications concerning physical reality—a pseudo-scientific doctrine and as such so much harder to kill.

37. *Uncertainty versus indeterminacy*

Let us examine the cornerstone of the Copenhagen interpretation, namely the Heisenberg uncertainty rule which, incidentally, is a special consequence of quantum mechanics rather than a principle in its own right. Heisenberg's relation, $\delta p \cdot \delta q \sim h$ ($\sim$ means about equal) describes the following experiment. Matter particles are directed toward a screen with a slit of width δq (or toward a solid strip of width δq in a reciprocal experiment). A film located at some distance behind the screen (or strip) records that the particles have spread out into a bundle of sidewise directions with sidewise momentum components p statistically distributed over a range $\delta p \sim h/\delta q$, conforming with quantum mechanics and in particular with the Duane-Epstein-Ehrenfest impulse transfer mechanism. The deflection of an individual particle cannot be predicted within the range δp; it remains *uncertain* in advance, though it can be ascertained *post factum*.

It is true that the same directional scatter can *also* be calculated from the hypothesis of (a) conversion of the particles near the slit into waves, (b) diffraction of the waves by interference, and (c) reformation of the waves into particles again, evasively called 'double manifestation'. But this uneconomical hypothesis has become expendable; quantum mechanics explains the statistical spread in a unitary way by momentum conservation. It must be emphasized therefore that Heisenberg's relation is *not* an argument for a dual nature of matter, first particles, then waves, then particles again, with or without the words 'as if' added. It merely delimits the uncertainty range for the *preparation and prediction* of an exact pair of q- and p-values for particles emerging from a screen with slit δq in space. (K. R. Popper([13]) 1935.)

Niels Bohr, however, wants us to replace Heisenberg's empirical result concerning uncertainty of prediction by an ideological indeterminacy of existence. He regards it as meaningless even to *think* that a particle within δq possesses a definite p-value within the range δp. In other words, electrons are not the kind of things which *have* exact position and momentum at the same time. I am slow to understand this. Coincidences in space and time, (q, t)-values, are the basis of all measurements in physics, other quantities being defined in terms of them. For example, two adjacent (q, t)-values define the momentum as $p = mv = m \cdot \delta q/\delta t$ in the limit of a differential quotient. Even before reaching the differential quotient dq/dt, the p-value at the place q is not meaningless, or not existing, or unthinkable. What can be measured also 'exists'. (This has nothing to do with the small radius of the electron.) Quantum physics deals with uncertainty of *pre-*

diction of what will be, rather than with non-existence or meaninglessness of what was or is.

How did Bohr ever come to declare an exact p-value of a particle within δq as meaningless? Because he first translated p into $h/\lambda = h\kappa$ of wave theory where it is indeed true that a wave number $\kappa = 1/\lambda$ of a wave aggregate condensed to a range δq is meaningless or undefinable within limits $\delta\kappa = 1/\delta q$; then translating the wave result into particle language again, he arrived at the meaninglessness of p within limits $\delta p = h/\delta q$. It is the same (faulty) reasoning which underlies the alleged duality dear to those unfamiliar with the third quantum rule of mechanics. Bohr's doctrine of indeterminacy of existence rests on a gratuitous literal transference of wave features to particles; it is a semantic artifice rather than legitimate physics. But after having been deluded by the duality idea to propose indeterminacy of existence, thereupon to maintain that the 'principle of indeterminacy' proves that particles do not have material existence, this is a feat of circular reasoning quite typical of the present quantum logic. In spite of having attracted much attention† it rests on the *salto mortale* of an arbitrary transference of wave qualities to particles in disregard of the statistical character of the quantum theory.

Several points in defense of indeterminacy of existence have been raised here. First, the momentum component p of a particle changes within the diffracting slit δq from one original to another final value. So how can one speak of the particle *having* one definite p-value when there are two? This objection is trivial; it would apply as well to the reflection of an ordinary billiard ball, without presenting us with a 'novel epistemological lesson' (Bohr's term). Furthermore, the argu-

† Quoting one among dozens of similar utterances: 'The suggestion has been made that the revision [of physics by indeterminacy] pierces to the basis of Man's fate and freedom, affecting even his conception of his capacity to control his own destiny. In no portion of physics does this suggestion show itself more pointedly than in the principle of indeterminacy of quantum mechanics.' From the authorized Introduction to Heisenberg's *Physics and Philosophy* by F. C. S. Northrop.

ment admits that the particle first *has* one and later *has* another p-value; it thus concedes what it pretends to deny.

Copenhagen's second line of defense runs as follows: The exact p-value emerging from δq cannot be measured directly within δq itself but only *indirectly*, by reconstruction of the path from the slit to the arrival point on the film. This objection is self-defeating, too. For how else can Heisenberg's rule $\delta p \sim h/\delta q$ be substantiated than from many arrival points of particles on the film? Denying a real meaningful existence of p-values within δp because they can be ascertained only by indirect means, yet using the same 'indirect' p-values and their spread on the film as testimony for a principle of indeterminacy of p-existence is circular logic again. Yet on this 'logic' rests the whole Copenhagen philosophy. Besides, nowhere in physics do we have 'direct' data, the only exception being location in space and time, that is (q, t)-values. Velocity, momentum, energy, etc., are always determined indirectly. Velocity, by its very definition, requires measuring two adjacent positions at two adjacent times. Mechanical and electromagnetic speedometers rely on theory which hardly is 'direct'. And a momentum measurement rests on the velocity measurement of a recoiling body. True, even though the particle emerges from the slit or its surroundings, it is not within the slit any more at the end of the measurement. Nevertheless, there is no point in denying that it must have come from the diffracting system with a definite sidewise momentum, rather than with a non-existent or meaningless momentum. I quite agree that the distinction between *being* and *being seen* which is irrelevant in dimensions of everyday life, becomes important in the microphysical domain where the act of being seen represents a relatively large encroachment upon the seen object. Nevertheless I cannot support a quantum school of thought which equates 'indirectly observed' with 'not observed', then with 'not observable', and finally with 'non-existent' and 'meaningless'. It is all too reminiscent of the sophisms of Zeno to whom it seemed meaningless that an arrow could *be* at a certain place and at the same time *change* its place.

It may be helpful to step down from the imaginary diffraction experiment illustrating the relation $\delta p \cdot \delta q \sim h$ to a concrete experiment illustrating the relation $\delta E \cdot \delta t \sim h$, namely the statistically ruled emission of α-particles from a radium preparation. Here δt is the uncertainty range of prediction of the time instant at which an individual α-particle will be emitted, the half-life of radium, $\delta t = 1580$ years. δt corresponds to a small statistical distribution range of the escape energies, $\delta E \sim h/\delta t$. According to the quantum logic above, however, one would have to maintain that the click in a Geiger counter announcing the arrival time of an individual α-particle constitutes only an indirect time measurement of its instant of escape, that direct escape instants do not really exist within the range δt, that they are meaningless—which to me seems a meaningless assertion indeed, proving the power of words over facts. Therefore I agree with Einstein's remark([9]): 'If one accepts the statistical interpretation . . . then the whole *egg-walking* in order to avoid the physically real becomes superfluous.' The 'Eiertanz' here is the transplanting of the age-old philosophical dispute about reality into the sphere of natural science where practical realism is indispensable and is safeguarded in atomic physics by the statistical theory of mechanical particle phenomena. In particular, Heisenberg's rules $\delta p \sim h/\delta q$ and $\delta E \sim h/\delta t$, describe an objective *statistical dispersion* of p- and E-values. The term *uncertainty rule* gives it a slightly subjective taint; but as a common experience it still belongs to physics. The assertions about *non-existence*, however, is of a purely metaphysical character; it certainly is not a lesson of modern physics.

Einstein has devised an imaginary experiment in which a direct measurement of an exact energy change δE of a body of mass m can be carried out at a time t without indeterminacy δt, by an instantaneous determination of its mass change, $\delta m = \delta E/c^2$. Bohr([10]) tried to refute him by the trump-card of Einstein's own general relativity theory of the gravitational redshift, that is, of the time retardation δt of a clock of mass m moved a distance δq against a gravitational field of potential gq, whereby the time shown by the clock suffers a relative retardation expressed by the formula

$$\delta t/t = g\ \delta q/c^2 = mg\ \delta q/mc^2 = \delta E/E.$$

The shifts δt, δq, and δE, are then taken as uncertainties of t, q, and E so as to vindicate the uncertainty connection between E and t. However, as J. Agassi remarked, if Bohr's considerations were conclusive, they would mean that the relativistic redshift could be *derived* as a consequence of quantum indeterminacy, or vice versa. This would be a most fundamental discovery if it were not so patently unsound. Indeed, how can one hope to justify the inverse proportionality of δE and δt of quantum theory by invoking the direct proportionality of the same quantities in the formula above? In contrast to the partisan view that 'all the shrewd objections raised by Einstein were successfully countered by Bohr' (L. Rosenfeld), I do not think that it is Einstein who has been refuted.

If energy measurements were restricted to the *indirect* Franck-Hertz optical method where the energy E of an incident particle is recognized by way of the frequency $\nu = E/h$ which it elicits from atoms, this frequency ν would indeed be defined only with accuracy $\delta \nu \sim 1/\delta t$, and E then could be found only with margin $\delta E \sim h/\delta t$. However, it is just Copenhagen's view that only *direct* measurements count.

The main difficulty for the idea that a particle is always at one point in space has been seen in the two-slit experiment. Here it is asked how a particle can contribute to the diffraction pattern unless it first transforms into a wave covering both slits simultaneously. However, as mentioned before, the way out of this dilemma is indicated by quantum mechanics, in particular by the Duane-Epstein-Ehrenfest impulse transfer mechanism: The diffraction pattern does not depend on the exact places where this or that particle reacts to the diffractor. The latter produces the deflection of the incident particles by way of its periodic components of matter distribution in space. Hence one linear particle path is as good as another for the observed statistical result, without electrons spreading out over both slits.

In conclusion: there are three interpretations of Heisenberg's relation, $\delta p \cdot \delta q \sim h$ (and $\delta E \cdot \delta t \sim h$):

(a) p and q cannot be exactly prepared and *predicted* simultaneously. This is the original uncertainty rule, a special part of quantum mechanics, confirmed experimentally.

(b) p and q cannot be *measured* simultaneously. This version is a half-truth. If one includes the word 'directly' then it is trivial because a momentum can never be measured directly. Without the word 'directly' it is wrong, since the value p acquired within δq *can* be determined, by reconstruction of space-time data and with the help of theory, which is the usual procedure in practical and theoretical physics.

(c) p and q have no exact coexistence; to speak of an exact simultaneous pair of qp-values is meaningless. This version works only when one arbitrarily replaces lack of predictability with lack of measurability, and then with non-existence.

Altogether, Heisenberg's rule neither deals with indeterminacy, nor is it a principle. It rather describes a special and most important example of unpredictability, when unpredictability as such was known as long as people placed bets on future events. Heisenberg's special rule follows from the general theorems of quantum mechanics which, in their turn, follow from still more general postulates of symmetry and invariance which alone deserve the name of 'principles'.

38. *Reconstruction of unpredictable data*

Closely related to the opinion that it is meaningless to think of qp-existence is the doctrine that one *can not know* what happens between two observational tests, hence it is best to think and say that *nothing happens*: 'If we ask, for instance, whether [between two observations] the position of the electron remains the same, we must say "no"; if we ask whether the electron's position changes with time, we must say "no"; if we ask whether the electron is at rest, we must say "no"; if we ask whether it is in motion we must say "no",' writes Robert Oppenheimer([11]). And Heisenberg ([7]) speaks of the 'hopeless difficulties to fill out what happens between two observations'. To me it seems just the opposite, namely that it is hopeless to describe what exactly happens *during* an observation, better: during the reaction of the object with a test instrument, whereas one often can fill out what happens *between* two observations. The latter is certainly true when the two observations are of the kind properly admitted as the only *direct* observations, namely coincidences in space and time, all other data being 'indirect' and containing some more or less disputable theory.

Take the example of a particle in a certain force field, observed first at the space-time point $(q_1\ t_1)$, then at $(q_2\ t_2)$. Irrespective of the height of the energy barrier between the two points, one can always reconstruct one and only one path connecting the two space-time positions according to the laws of classical mechanics. Indeed, the first measurement affects the energy and momentum vector of the particle in an unpredictable manner. The second (q, t)-measurement then reveals which E and p the particle had actually acquired during the first (q, t)-measurement so as to bring it from the first to the second point on a continuous path in the required time interval. I therefore submit that for the standard case of space-time location measurements (coincidences) *one always can fill out* the gap by a definite path with definite locations at all intermediate instants.

Denying that a particle always *is* somewhere is not even warranted by diffraction experiments since, according to quantum mechanics (in contrast to the wave manifestation story of three decades ago) each particle reacts to a space-extended periodic component in the matter distribution of the diffractor; it is irrelevant for the deflection where exactly the reaction takes place. But saying that it takes place nowhere, that it is meaningless to assign some, though unknown, position to a particle at all times, is not warranted, neither experimentally nor theoretically. It is a linguistic extravaganza rather than a philosophical innovation.

Difficulties of reconstruction occur in case of *indirect* quantities. The simplest case is that of a velocity, defined as the differential quotient $v = dq/dt$, approximated experimentally by the quotient $(q_1 - q_2)/(t_1 - t_2)$ for adjacent points. There are no direct methods for determining a velocity. Speedometer readings are certainly indirect; they rest on centrifugal or electromagnetic theory. Moreover, as Margenau emphasized, velocity, momentum, and other indirect quantities allow several operational definitions, none of them having a claim to being the only correct one. Take the case of a free particle running forth and back along a line of length L. Suppose the particle is observed at two space-time points, 1 and 2. According to one definition, the velocity between the two measurements is $(q_1 - q_2)/(t_1 - t_2)$, whereas according to the quantum theory the velocity in this arrangement can only have value $v = p/m = h/2mL$ or a multiple thereof. It is of course the easiest way out of this conflict to decree that the particle *has no* velocity between the two (q, t)-observations, and that it is meaningless even to ask whether it moves. A similar dilemma may occur for other 'indirect' observables. But saying that nothing happens is neither confirmable nor falsifiable by experiment; it is an empty phrase. The root of the difficulty of reconstructing values of indirect observables may be seen in an ambiguity of their definition which always requires theory. For example, the two definitions of the kinetic energy, $p^2/2m$ and $pqp/2qm$, are identical in classical mechanics. But they differ in quantum mechanics where p is the differential operator $(h/2i\pi)d/dq$ which does not commute with q; so which definition is right?

The distinction between direct and indirect observables

must not be confused with that between primary and secondary qualities of eighteenth-century empiricism which regarded location in space and time as a primary objective quality, in contrast to color, odor, taste as secondary qualities
which vary from subject to subject. The question then was:
'What can we know with objective certainty about things?' In
physics, however, we ask: 'What can we measure with instruments?' irrespective, of course, of whether we or others acquire
conscious knowledge of the data. I here agree with R. D.
Bradley[12] when he wrote: 'The notion that an atom has no
properties at all when it is not observed is the supreme
example of a purely metaphysical thesis masquerading as a
physical theory.' Refer also to K. R. Popper.[13]

A problematic situation connected with indirect observables
occurs in the following example. According to the Schrödinger
equation, a particle of energy E_1 has a finite probability to be
found at places q_2 which would require it to have a larger
energy $E_2 > E_1$. How can a particle ever 'tunnel through to
an energetically unaccessible point'? Quantum mechanics
answers that the particle, by virtue of its reaction to the position meter ascertaining q_2 may have *changed* its energy within
a wide range δE including the required energy E_2. It certainly is not justified to exclaim: 'You see that matter has a
wavelike nature since the forbidden point can be reached only
under the wave equation.' A wave equation indeed; but this
does not imply that electrons are 'wavicles'. Schrödinger's
equation does not deal with matter waves but with probability
amplitudes. We cannot accept the statistical meaning of ψ outside the tunnel where it has been confirmed experimentally,
and forsake it inside the tunnel.†

Let us distinguish therefore between what is broadly called
the Copenhagen language, abbreviated C, in contrast to a more
realistic view, R. According to C no exact p-value exists in,

† These considerations may show that I have come much closer to the
realism plus empiricism postulated by H. Mehlberg in his 'Comments',
p. 360-70 in *Current Issues in the Philosophy of Science*, edited by H. Feigl and
G. Maxwell, New York, 1961.

and emerges from, δq; according to R an exact value p exists within δq and can be reconstructed from later observation. According to C a particle *is* nowhere between two observations of its position, nor does it *have* an exact E-value unless E is certified by direct observation. (I do not know what constitutes a 'direct' energy observation); according to R, a particle is always somewhere and can have energy and position at the same time in spite of the 'incompatibility' of these observables, that is the impossibility of preparing and *predicting* exact (E, q)-pairs. There are exceptional cases, for example the tunnel effect, where this simultaneity of existence leads to difficulties according to classical ideas; they do not warrant, however, a denial of coexistence in all cases. According to C, an α-particle emitted with energy margin δE does not *have* an exact time instant of escape; according to R it has a definite escape instant, although the latter cannot be predicted or prepared. According to C, the spin axis of a particle before entering a magnetic field does not *have* a direction at all; according to R, a spin direction exists even before entering the field.

39. *The mystique of wave packet contraction*

It would be unfair to quote views enunciated during the first turbulent years following the establishment of the quantum formalism in 1926, if these views were not still presented as modern science. Thus, in his Discussion with Einstein of the diffraction experiment (1949) Bohr [10] speaks of 'the wave connected with the motion of the particle'. Here we must ask: What sort of wave? A pilot wave spread out in space guiding the particle, as de Broglie has it? Or a mental wave of absence of knowledge? The answer may be contained in Heisenberg's description of 1930 of what happens when a particle (whether of light or of matter is irrelevant here) is found on the reflected side of a semi-reflecting boundary when it could as well have been transmitted to the other side [2]: 'The experiment [of catching the particle] at the position of the reflected wave packet exerts a kind of action, a reduction of the wave packet, at the distant point occupied by the transmitted wave packet; and one sees that this action is propagated with a velocity larger than that of light.' Heisenberg then assures us that the 'action' exerted by the experiment at one place, and transmitted with super-luminal velocity to another place, does not violate the theory of relativity since it cannot be used to transmit a signal. Fair enough! But according to the statistical interpretation, a ψ-function is a probability table, not essentially different from any mortality table. Suppose that such a table is used in the city A to insure a man on a future trip allowing him to travel East and another table for the West. Later he happens to die at a place W west of A. Do the mortality tables collapse with supra-luminal velocity by a 'kind of action' from W to all points east of A? This was in 1930, three years after Born's statistical interpretation brought clarity.

Today there is a more subtle version of the wave packet contraction story: The two halves of the wave packet do not actually run along the two sides of the semi-reflecting boundary through space. They are only mental pictures. The mind picture of the two wave packets instantaneously contracts when the particle (which also is a picture, dual to the wave picture) is recorded on the west side, or according to the very latest version, only when the record enters the conscious knowledge of an observer—if I understand this psycho-physical theory correctly. But one never knows whether ψ describes waves in physical space associated with particles guided by the Schrödinger equation, or whether ψ is a mental picture of possible actions in the particle picture. We therefore turn for advice to Heisenberg again:

'It is entirely possible to imagine this transition, from the possible to the actual, moved to an earlier moment of time.'

To those who wonder why the transition can be moved to an earlier time, Heisenberg answers: 'For the observer does not produce the transition.' This is rather puzzling as long as one does not know which part of the transition from the possible to the actual is mental and which is physical. But Heisenberg answers it all by the following declaration:([7]) 'The representation of a packet of probabilities is completely "objective", that is it does not contain features connected with the observer's knowledge. But it also is completely abstract and incomprehensible since the various mathematical expressions $\psi(q)$, $\psi(p)$, etc., do not refer to a real property. It so to speak contains no physics at all.' This is Hegelian dialectics in reverse. According to Hegel, a mathematical concept does not exist but is real, whereas an individual thing exists but is not real. To Heisenberg the mathematical concept of a ψ-function is objective but not physical, whereas a thing, a particle, is physical but not objective, being no more than a subjective picture. Borrowing a passage from James R. Newman applied to another case: 'One is apt to agree with what he says, without knowing exactly with what one is agreeing.' Indeed, when 'objective' goes together with 'abstract and incomprehensible', with 'no real property' and 'no physics at all', and when the beginning of 'the actual' can be moved to an earlier moment of time, one hesitates to agree to such scholastic play with philosophical vocabulary. Also, when Heisenberg invokes the *potentia* of Aristotle in connection with the uncertain outcome of an atomic test, one wonders why Aristotle's *potentia* are not also needed for clarifying the uncertainty in ordinary games of chance. And Heisenberg's remark that 'the reduction of wave packets cannot be derived from Schrödinger's equation' is trivial unless he still flirts with the idea of an 'action propagated with supra-luminal velocity through space'. Nor can I understand why a particle, after reacting to a semi-reflecting boundary, *is* neither on the right nor *is* it on the left of the boundary until a 'transformation from the possible to the actual' has entered someone's consciousness. With due respect, it seems that Heisenberg merely describes in exotic terms what can as well be told after re-entry from the clouds in a more earthbound idiom as follows:

A micro-object in the state S_m, as revealed by an S-meter, may next be tested by a T-meter and thereby be thrown into a state T_n. And this is all; further talk about the 'abstract and incomprehensible' and other philosophical subtleties can only promote confusion. It is true that the original state S_m is connected by a probability function with the various states $T_1 T_2$ . . . which might evolve in a T-test. But there is no reason to attribute to this probability function, to this 'wave packet', Heisenberg's mystical features quoted above. But von Weizsäcker, inspired by the ψ-philosophy, cuts the Gordian knot of profundity by the following declaration([1]): 'We can best speak of a collapse of the category of substance . . . or rather of adapting our logic, formed by thinking in objects, to the new situation.' Although this seems pregnant with cosmic perspectives, almost as far-reaching as 'piercing to the basis of Man's fate and freedom' (see p. 120) I am again slow to comprehend. First, it is somewhat heroic to conclude, because a q-measurement involves an unpredictable p-recoil, that the category of substance has collapsed. Second, logic is a purely formal discipline and has nothing to do with whether one 'thinks in objects'. Third, applying three-valued logic of Yes, or No, or Undecided, is illogical because the Yes and No refers to *facts*, whereas the Undecided concerns to what someone may *think* about facts which have not yet happened: False opposites again! Fourth, every statistical pattern, even the Gaussian in case of geometrical symmetry, is a sort of puzzle for reasons discussed in Chapter II. But I cannot bring myself to adapt my logic, not think in objects any more, etc., when besides the Gaussian there also are alternating maxima and minima.

Von Weizsäcker's statement above is symptomatic of a state of mind, endemic among quantum interpreters, which the logician Wittgenstein has diagnosed as 'mental cramps arising from linguistically created puzzles'. Indeed, committing the original sin of accepting the word 'duality' as an explanation is returning to the nominalism of the scholastic age.

40. *Subjective interpretation*

Skepticism about 'thinking in objects' is an old story. Every ostrich and every babe is a Berkeleyan idealist to whom things exist only when seen, and discontinue to exist when covering his eyes. But common sense and science alike are based on the opposite idea. Even when certain qualities of an object change by looking or measuring them (the momentum of an electron changes unpredictably by its recoil to light reflection), there are other permanent qualities, mass, charge, spin, which characterize a *thing* in an objective way. There are no grounds for not 'thinking in objects', let alone 'adapting our logic' unless one confuses variable qualities with *things* actually characterized by permanent qualities. When the energy of an electron in a H-atom has been ascertained, indirectly of course, then one cannot predict the result of a subsequent position measurement at a certain time; this is a physical fact. But maintaining that the electron *has* no position at all between two measurements is rhetoric, rather than a revolution in the theory of knowledge. All this has been said by others before. Refer to the 1935 criticism by K. R. Popper[13] and to the more recent writings of M. Bunge[14], P. Feyerabend[15], and H. Margenau [16], to name but a few representatives of the rising flood of opposition to the Brave New World constructed in the Danish capital and surrounding feodalities out of the great advances in physics achieved at the same places.

I know of course that quantum theory has introduced very important innovations into the science of mechanics. The novelties do not consist, however, in the discovery that there are mutually incompatible observables as such. Plenty of ordinary examples can be mentioned to illustrate the same concept. Nor does the novelty consist in the realization that *atomic* statistical distribution defies causal explanation; the same holds for every statistical distribution in any game of chance. The special lessons of quantum physics may rather be summarized as follows:

132

(1) There is a class of 'observables' $A, B, C, \ldots$, pertaining to a given mechanical system, capable of certain sets of 'eigenvalues', $A_1 A_2 A_3 \ldots, B_1 B_2 B_3 \ldots$, and so forth. They are ascertainable by means of A-meters, B-meters, etc. mostly in a very indirect fashion. The eigenvalues belong to corresponding 'states'.

(2) The 'states' are connected by probabilities such as $P(A_k, B_j)$ which are of a two-way *symmetry*, leading to the result that the P's can be arranged in matrices, all of the same multiplicity, their M rows and M columns adding up to *unity*.

(3) The postulate that the P-matrices be interdependent by way of a *general*, that is symmetric and transitive theorem is satisfied only by unitary transformation, isomorphic with the interference of probabilities by way of probability amplitudes ψ.

(4) The postulate that the probabilistic dynamics be invariant with respect to linear shifts of zero points in q- and p-space entails that the probability amplitude function $\psi(q, p)$ is of the periodic form $\exp(2i\pi qp/\text{const})$, the constant being known as Planck's h.

These features characterize quantum mechanics within the general category of games of chance. They cover all the wave qualities attributed to ψ, the non-commutativity of factors in symbolic products, and all the rest. There is nothing incomprehensible and abstract which would call for a new theory of knowledge in which the category of substance breaks down, and where we have to 'adapt our logic'. Quantum mechanics describes relations between data of macroscopic instruments. I do not say 'readings' in order to avoid any allusion to reading subjects. Knowledge and conscious reading by human observers are as irrelevant in atomic physics as they are in any other branch of physical science. To speak of the contraction of a wave packet upon an observation is as senseless as to speak of the sudden contraction of a mortality table upon an individual fatality. It is a piece of antique furniture which ought to be discarded in the interest of clarity, the sooner the better.

41. *Measurement*

In order to measure the height of an energy level, a position at a certain time, or any other observable named A pertaining to a certain object (supposing we still 'think in objects', disregarding von Weizsäcker's admonition), we need an instrument, an A-meter which, in its turn, defines what is meant by the observable A. Speaking in general terms, the pointer position along the A-scale then will indicate a certain value A_k. Atomic measurements usually require a large amplification, such as droplet condensation or bubbles, blackening of photographic grains, avalanches in Geiger counters, and the like. Such measurements are always indirect. Only idealized experiments may be imagined as giving 'direct' evidence, whatever that means. Concerning the significance of a measuring result, there is a difference between classical and quantum theory, however. According to classical ideas, the value A_k can be attributed to the object immediately before, during, and after the A-measurement. According to quantum physics, there is an active, unpredictable, and unavoidable participation of the A-meter in producing the result, by throwing the microphysical object from its previously occupied state S_m determined by an S-meter into the new state A_k. Therefore, the measured value can be ascribed to the atomic object only immediately *after* the measurement is completed. A subsequent B-meter yielding the value B_n wipes out all traces of the previous state A_k and produces an entirely new situation. By the way, in the standard case of two space-time location measurements one can reconstruct a connecting path according to classical mechanics, as discussed in section 38.

Trouble begins, however, when the material body used as a measuring instrument is regarded as a *subject*. And when it is said that quantum theory has changed the relation between subject and object, this makes a great impression on those who are led to identify statistical distribution, recorded by instruments, with the knowledge or lack of knowledge of observing subjects. The subjectivistic trend began with the two mental pictures in place of one physical reality of matter particles. And it reaches its peak when *human consciousness* is presented as an essential element of atomic theory, for example when E. Wigner declared at a recent Symposium on Physics and Philosophy (Marquette University, 1961): 'Physicists found it impossible to give a satisfactory description of atomic phenomena without reference to the consciousness. . . . An atomic measurement is not completed until its result enters the consciousness.' With due respect, I find it impossible to understand why the velocity measurement of an electron between two charged plates can be described only with reference to the consciousness, whereas the velocity measurement of a falling stone can do without consciousness. Or is consciousness needed only in a statistical contingency where the result cannot be predicted? If so, why must the unpredictable reaction of an electron going either to the right or left in a magnetic splitting experiment (Stern-Gerlach) enter the consciousness in order to be completed, whereas the same is not necessary for a ball falling to the right or left of a blade? Even more am I mystified by the often repeated phrase that an electron in the Stern-Gerlach two-way deflection experiment *is* neither on the right nor *is* it on the left until its record of arriving here or there has entered someone's consciousness. To say the same of the ball in the ball-blade game would not make sense, of course. But since the difference between electron and ball is one of size, one would like to know whether consciousness enters gradually via semiconsciousness, or abruptly at some critical size. I am not the only one who holds that quantum physics, experiment as well as theory, deals with records of instruments rather than with observers' consciousness, with objects rather than with mental pictures, with statistical distribution rather with lack of knowledge of human observers. Whether the record of an instrument is 'read' by this or that person who takes conscious notice of it is absolutely, positively, and emphatically irrelevant. I am entirely on Reichenbach's([17]) side when he wrote: 'Like all other parts of physics, quantum mechanics deals with nothing but relations between physical things; all its statements can be made without reference to an observer. The disturbance by means of observation—which certainly is one of the basic facts asserted by quantum mechanics—is an entirely physical affair which does not include any reference to effects emanating from human beings as observers. . . . The instrument of measurement disturbs, not because it is an instrument used by human observers, but because it is a physical thing like all other physical things.'

I also follow Margenau([18]) when he asked 'whether physics has seriously begun to describe human knowledge, a subjective aspect of the mind, in terms of differential equations involving physical constants. The point I wish to make is that we are not forced to this conclusion', with which I emphatically agree. Quantum theory deals with objective situations, and not with packets of expectation spreading out in space and suddenly collapsing as though through a kind of telepathy. The quantum theorist, as every scientist, coordinates data recorded by (macrophysical) instruments, data which are 'objective', that is *reproducible* by immediate repetition of the same test with the same object, although it may often be difficult or impossible to recognize 'the same atom' in the 'same state' by means of the 'same instrument'.

It is true that the outcome, the passing or not passing of a particle in an individual test, is not uniquely determined by its previous state but is ruled statistically. Still, subjectively tainted expressions such as uncertainty, probability, expectation, etc., signify statistical ratios of *objective data* recorded in many actual experiments, collected in tables or represented as functions. The fact that the physicist takes statistical results as a topic for his personal reflection, calculation, expectation, and as betting odds for future tests, etc., is unessential; it is not even characteristic of quantum theory. Any classical law, such as $s = \frac{1}{2}at^2$, is first a generalized condensation of past experience; it also can be used for planning future experiments. Whether the odds of expectation are spread statistically over a wide margin, or condensed in one value with certainty, is a matter of degree, though of vital importance to the physicist. Still, quantum theory, like classical theory, connects data recorded by instruments. Speaking of wave packets of expectation in space, or in an observer's brain, is as tortuous as speaking of six threads of expectation in a dice game which suddenly break when a die is cast, or is seen by an observer to be cast. Notions like this are left-overs from the heyday of the theory when one believed in ψ-functions describing physical states of wavelike vibration. Actually, ψ-functions are tables of betting odds provided with vector directions so as to fit them into the schema of unitary transformation.

But why is there so much discussion about the meaning of an atomic measurement? The reason, it seems, is the erroneous connection of a physical datum A_k (displayed by an A-meter) with a set of *possible* future data $B_1 B_2 \ldots$ to be found in another measurement with a B-meter, described by a function $\psi(A_k, B_j)$ where B may be any other observable whatsoever. When ψ is used as mathematical representation of the one physical state A_k there is no harm. But when one particular function $\psi(A_k; q, t)$ is regarded as essential for the characterization of the state A_k, the door to misunderstanding is open, the more so as there are at least seven meanings attributed to this particular ψ, listed in Section 42. One of them is the myth that ψ describes a physical state of matter spread out in space and time (or in someone's thinking about space and time) which suddenly contracts to one point when the particle is recorded at one place (q, t), or perhaps only when an observer becomes conscious of this recording.

But let us hear what Heisenberg thinks about an atomic measurement as far as can be gathered from his writings:

(a) Before the measurement the state of the system is represented by a function $\psi_0(q, t)$. This ψ_0 changes with time t

according to the Schrödinger wave equation as a 'process equation' describing a gradual change of state.

(b) During the measuring act, the state changes suddenly, resulting in a new function $\psi_1(q, t)$. This sudden change is denoted as a wave packet contraction, not controlled by the Schrödinger equation.

(c) After the measurement is completed (by entering the consciousness of an observer), the state changes continuously again, the state function ψ_1 being subjected to the Schrödinger equation.

Opinions are divided as to whether the 'contraction of a wave packet' takes place with the velocity of thought as a mental process, or with large velocity through space as a physical process. Heisenberg seems to lean to the latter view: he assures us that, in spite of the supra-luminal velocity, relativity theory is not violated since the process of wave packet contraction cannot be used to transmit a signal from one place to another. On the other hand, he leaves it to our choice when the Schrödinger equation is to stop working, either when an instrument records a new datum, or only when an observer takes conscious notice of it.

Fortunately, quantum mechanics which, in its physical development, owes so much to Heisenberg, tells a different story:

(a') The object may be in the state S_i in which it emerged from an S-meter. The one state S_i, as a vector in the space of unitary transformation, can mathematically be represented by the set of its components $\psi(S_i, A_k)$ with respect to the orthogonal set of states $A_1 A_2 \ldots$ The same state S_i can also be represented mathematically by its components $\psi(S_i, B_j)$, and so forth. Each set of components may be written as a function, $\psi(S_i, A)$ or $\psi(S_1, B)$, and so forth. All these functions represent the one state S_i as a vector in Hilbert space.

(b') When an R-measurement is carried out, the object emerges in one of the states R_n with probability $P(S_i, R_n) = |\psi(S_i, R_n)|^2$. The new state R_n can again, as a vector, be represented by the set ψ of its components to any other orthogonal set of states. Further comment ought to be based on the quantum mechanics (a') (b') in which the word 'state' has one meaning, physical state of the object as ascertained by a test instrument. In (a) (b) (c), however, 'state' is used in two different meanings, physical state of the object, and state of knowledge by an observer. A measurement can only ascertain an objective state, acquired under the influence of the instrument, the R-meter, which produces one of the states R_n in an unpredictable fashion. What happens between two measurements, S_t and R_n, can often be reconstructed afterwards with the help of this or that theory.

42. *Seven meanings of* ψ

It is small wonder that misunderstandings about the role of the ψ-function with its alleged sudden contraction, with its half physical, half psychological content, etc., has led to the widespread view that the quantum theory, and the complex quantity ψ in particular, is abstract, sophisticated, fictitious, unpictorial, incomprehensible, has no direct physical meaning, and yet for some unaccountable reasons leads to correct results. I think that the simple outline of the theory developed in the previous chapters does not support this ancient folklore any more.

Indicative of the present confusion regarding ψ is the following list of seven opinions as to what a ψ-function describes:

(1) The physical state of a continuous material medium in space and time, with particles as mere appearances (Schrödinger 1926).
(2) A continuous pilot wave which controls point events along its course (de Broglie).
(3) A fluid containing hypothetical quantum forces invented *ad hoc* so as to determine its own motion according to the laws of hydrodynamics (Bohm).
(4) *One* definite state, or a sequence of *many* states of an object, or of a statistical ensemble of objects (the textbooks vacillate between these opinions).
(5) A wave state in space (for example in the transmitted as well as in the reflected part of a 'wave packet' in case of partial reflection) contracting with super-luminal velocity either when a point event takes place (in one of the two parts) or only when an observer gains knowledge thereof (Heisenberg, 'waves of expectation' as psychic actualities).

(6) A mathematical symbol incapable of pictorial representation, completely abstract and containing, so to speak, no physics at all, yet completely 'objective' since not referring to any observers' knowledge (Heisenberg, the Copenhagen language).

(7) A ψ-function is a well-ordered list of betting odds, based on statistical experience, for the diverse outcomes of specific tests of a microscopic object with a macroscopic instrument.

However, although it may be good for politics to 'let a hundred flowers bloom and let a hundred schools contend' (Mao Tse Tung), it is not good enough for science; here only the interpretation (7) stands up to realistic criticism in accordance with monolithic quantum mechanics.

A great deal of confusion has its origin in the early, but wrong, analogy between matter particles and photons on the one hand, ψ-waves and electromagnetic waves on the other. There are individually identifiable matter particles (producing tracks, etc.), but there are no 'photons' as real identifiable things as discussed in Section 36. On the other hand, electromagnetic fields are real (kicking) things, whereas ψ-waves are tables of probabilities for events (rather than real things as Schrödinger first thought). The quantum theory can be applied to real things, to matter particles as well as to electromagnetic waves. In particular, if the motion of a system is harmonically analyzed into components of frequency ν_k, then quantum theory says that it can give out and receive energies only in amounts $h\nu_k = E' - E''$. Similarly, an electromagnetic or other field may be analyzed harmonically into standing vibrations of various frequencies ν_k which change their energy only in quantized amounts $h\nu_k$ in resonance with a corresponding change of a matter particle. However, this change of field energy must not be thought to occur locally at the place of the particle, as though a 'photon' were swallowed or emitted. The change of radiation energy rather takes place in the standing wave as a whole. This at least is the view of the *quantum theory of radiation* as developed around 1930 by Fermi, Heitler, Dirac, and others. There are no localized photons. And quantum theory regards the creation and annihilation of matter particles not as true creation out of nothing or annihilation into nothing, but as transitions of the particle from positive to negative energies, according to Dirac's hole theory; the particle thus persists as a 'thing'.

Insurance companies have permanent tables for statistical odds under various physical conditions; similarly, quantum theorists have P- and ψ-tables and functions. But why such a table or function should suddenly collapse or contract (Heisenberg-von Neumann 'picture', Section 39) upon an actual event is beyond the writer's capacity to grasp, in quantum physics as well as in insurance. What 'happens' is that persons interested in betting on future measuring results may turn to this or that new page in their catalogue of P-tables or ψ-functions. But the turning to a new page, wrongly dubbed 'contraction of a wave packet', has of course nothing to do with any actual physical process. Yet, in order to give to the 'contraction' a semblance of respectability, *ambiguity of word meanings* is applied: To Heisenberg 'state' first means physical state of an atom, afterwards state of knowledge of an observing subject. In preparation for this ambiguity, Heisenberg tells us that already the classical concept of temperature sometimes describes the physical state of a system, on other occasions the state of knowledge or absence of knowledge of an observer. He regards this iridescence (das Schillern) of word meanings as an essential feature of every statistical theory. In the same tenor he remarks([19]): 'When the old adage "natura non facit saltus" is used as a basis for criticism of the quantum theory, we can reply that certainly our knowledge can change suddenly, and that this fact justifies the term "quantum jump".' Is there really no better way to justify the term 'quantum jump' than by referring to sudden changes of personal knowledge? At a recent Symposium on 'Observation and Interpretation' one discussant justly remarked: 'Nobody will deny that the laws of quantum physics were valid already at the time of the Saurians. If so, then they are objective laws which are independent of human observers and their knowledge.' Whereupon another member retorted: 'But they could only be *known* by observers.' Since the same remark can as well be made with respect to any classical law, and with respect to anything that happens anywhere in the world, it has absolutely no relevance to quantum as opposed to classical physics. It only reveals the complete absence of judgment as to what matters and what does not matter, as soon as people enter the labyrinth of present-day 'quantum philosophy'. The same lack of sense is shown when another famous interpreter, Sir James Jeans, told his spellbound readers: 'It is probably as meaningless to discuss how much room an electron takes up, as it is to discuss how much room a fear, an anxiety, or an uncertainty takes up.' Such utterances call for massive retaliation.

43. *Bohm's sub-quantum theory*

When balls are dropped along a board spiked with nails (*Galton* board) they arrive at the bottom with a statistical distribution known as the Gaussian law. This law can be deduced by pure mathematical reasoning from the assumption that each ball has an even chance of dropping to the right or left of a nail. The Gaussian distribution is thus reduced to simple fundamentals; it does not require any further explanation. However, someone proposes the following 'causal' theory. First he symbolizes the Gaussian distribution as a gas. Then he wonders why his self-invented 'gas' displays a density maximum in the centre rather than spreading out evenly, as a real gas would do. Third, he embarks on inventing a hydrostatic force distribution throughout his fancied gas adjusted in such a manner that, *if* this force existed, it would produce the same density distribution in a real gas according to the gas pressure laws which the 50:50 probability yields anyway according to *Gauss*. An *ad hoc* constructed hypothetical force in a non-existent gas may have esthetic and even heuristic value. Regarding it as a causal explanation of Gauss's law on a deeper level of understanding would stretch the meaning of the words 'explanation' and 'understanding' beyond recognition, however.

It so happens that *Bohm's* theory is the exact analogue of the above 'causal explanation' of the Gaussian distribution. Bohm first reinterprets the statistical density $|\psi(q, t)|^2$ in *potential* experiments as the density of a fluid. Then wondering why his self-invented fluid is not distributed with constant density in space and time, as would an ordinary fluid in the absence of compressive forces, Bohm goes on inventing a hydrodynamic force distributed throughout his background fluid distributed in such a manner that, *if* it actually existed, it would produce the same space-time distribution according to gas dynamics which the fundamental probability metric, and the Schrödinger equation as a special application of this metric, yields anyway. Finally he would have us believe that the individual particle positions are floating along in the stream of this background fluid.

There are three fundamental objections to this theory. First, following wave mechanics, Bohm's picture suggests that $\psi(q, t)$ represents an actual statistical distribution state of particles over q at various times t. In fact, $\psi(q, t)$ stands for $\psi(S; q, t)$ indicating that the particles *are* all in the same state S, and that $|\psi(q, t)|^2$ is their probability of jumping from S to q at t if someone *should* carry out q-tests (rather than A-tests in general, q and t having no special preference in the general probability interference metric). Second, a properly adjusted 'quantum force' invented *ad hoc* hardly offer a deeper understanding of the particle behavior. It would merely be an interesting *analogy* to the laws of hydrodynamics, but rest on no more fundamental grounds than a hydrostatic 'explanation' of the Gaussian distribution at the bottom of a Galton board. Third, Bohm violates the symmetry of q and p which is essential for mechanics, classical as well as quantal.

It is hard to see, then, how far Bohm's attempt of reducing simple and elementary laws of probability to a complicated *ad hoc* invented deterministic substratum should contribute to a better understanding of the statistical behavior of atomic particles. 'Hidden parameters' can always be invented at will, and their causal action can then be explained by another set of still more hidden parameters, without need for experimental confirmation just because they are 'hidden'. If there is jubilation in certain quarters that 'Bohm has shown that a causal interpretation of quantum theory is possible after all', we can only reply that his 'causal explanation' violates the standards of scientific method: 'Do not try to explain simple results by *ad hoc* invented complicated hypotheses.'

On the other hand, my special reason for regarding Bohm's search for a sub-quantum theory as futile is the undeniable fact that the quantum formalism, which once looked so arbitrary and *ad hoc* invented, turns out to be anchored in a non-quantal basis of great generality and simplicity which can hardly be surpassed. The postulate of invariance with respect to shifts of zero points would have to be maintained even in the vague plan of atomizing time and space themselves, whatever that means. And without the symmetry, $P(A \rightarrow B) = P(B \rightarrow A)$, there would be no equilibrium states; the whole of statistical thermodynamics would crash down in ruins. What else is there to modify? It seems to this writer that the developments of the preceding chapters furnish the very sub-quantal level for which Bohm and his supporters are searching. The objects (particles, fields) to which the quantum rules are applied may change, but not the formal quantum rules as long as the underlying non-quantal postulates apply.

44. *Bohr's universal complementarity*

One of the foremost achievements of quantum physics has been the discovery that coordinates q and momenta p are mutually incompatible, which is to mean that one cannot prepare and *predict* any simultaneous qp-states. This is certainly of great physical interest—as interesting as the opposite result that energy E, angular momentum A, and z-component of A are compatible, that is that there are reproducible

EAZ-combination states. But why is it of great *philosophical* significance that some groups of observables are compatible, and others are not? Flexibility and optical transparency data of a celluloid plate are compatible, flexibility and combustion heat data are not. This is an important technical detail, but it does not affect the theory of knowledge, as *qp*-incompatibility is supposed to do. Nor is there reason for changing 'The World View of Physics' when various probabilities are found to be connected by a *law*, and when this law of 'unitary transformation' between unit magic *P*-squares is best expressed in terms of a geometrical law with each *P* having a *vector* direction in a plane structure, denoted by the complex symbol ψ. Therefore I cannot see that 'the logical(?) structure of quantum mechanics has made profound changes in our scientific thinking' (Bohr). Flexibility and combustion heat data of plastic materials are incompatible, too. Yet their incompatibility has never made profound changes in our scientific thinking. Nor is our *thinking* changed when micro-physical experiments are found to have the character of (ordinary) games of chance—except under the erroneous notion that quantal statistical distribution can never be reduced to deterministic interpretation whereas classical statistical distributions can. Actually, the difference between an ordinary game of chance and the atomic *pq*-game is technical rather than essential. The *pq*-game is a physical innovation but it does not give us an epistemological lesson.

One of the most creative and critical spirits among theoretical physicists, Fritz London, wrote already in 1937 to the author (who at that time worshipped complementarity): 'To be frank, I do not have much taste for the complementarity of the wave and particle picture at all. As I see it, these are crutches which originally helped us to become familiar with quantum ideas, but do not describe the present situation any more.' The prolonged effort to save dualism and complementarity by maintaining that neither 'picture' is true, has resulted in an almost total eclipse of the intellectual curiosity to find

out *why* particles produce wavelike statistical appearances.
Instead of laying bare the deeper correlations which lead to
dualistic appearances, the primary goal has been that of de-
fending and then enlarging the doctrine of duality and complementarity. Niels Bohr in particular believes to have found
confirmations of complementarity not only in atomic physics
but in other fields of human knowledge, in biology,
psychology, sociology, and so forth. 'We may truly say that
different human cultures are complementary to each other',
further examples being 'justice versus charity' and 'the individual versus his community'. No wonder if others wish to
participate in this movement, for example when Jean-Paul
Sartre[20] sees complementarity also in the eternal contrast of
male and female; he takes it for granted that we know which
is waves and which is particles. And Bohr's close associate,
J. A. Wheeler, exclaims: 'Complementarity, in one sense,
represents the most revolutionary conception of our day.' In
this general sense I cannot agree, particularly in view of
K. R. Popper's poignant remarks pertaining to the analogous case of the all-embracing theories of Freud, Adler,
and Jung: 'It was precisely this fact, that they always fitted,
that they were always confirmed, which in the eyes of their
admirers constituted the strongest argument in favor of these
theories. It began to dawn on me that this apparent strength
was in fact their greatest weakness.' But returning to physics
again, we are told by L. Rosenfeld[21]: 'The complementarity
idea is, first of all, the most direct expression of a fact . . .
as the only rational interpretation of quantum mechanics.'
To me the contrast between a thing and one of its qualities
is irrational rather than 'the only rational interpretation'.
But Heisenberg[19] goes even farther and declares:
'The concept of complementarity . . . has encouraged
the physicists to use ambiguous language, to use classical
concepts in a somewhat vague manner in conformity with
the principle of uncertainty.'

Although the concept of complementarity may indeed have

encouraged ambiguous language, I cannot see much merit in
it. In the contrary, it would seem that a doctrine leading to
ambiguous language ought to be scrutinized for flaws of logic.
Heisenberg, however, continues: 'When this vague and unsystematic language leads to difficulties, the physicist has to
withdraw into the mathematical schema and its unambiguous
correlation with the experiment.' Again I would think that
the physicist does not need to withdraw unless he has used
vague and unsystematic language in the first place, that is
unless he has succumbed to the chronic illness of confusing
lack of predictability (which is physics) with non-existence
(metaphysics), uses illogical contrasts of incontrastibles, indulges in wrong physics by ignoring the third quantum rule
although applying the first two, and all the rest. When
Heisenberg finally generalizes: 'Insistence on the postulate of
complete logical clarification would make science impossible,'
then even the faithful will pause and wonder. They may
be tempted to listen to the sociological analysis of Chr.
Caudwell in *The Crisis in Physics* ([22]): 'The philosophy of
all bourgeois philosophers in relation to matter is the same:
mechanical materialism. But for various historical reasons
bourgeois philosophers ceased to be interested in matter and
developed another part of bourgeois philosophy, that concerned with the mind or subjective reality.'

In contrast, L. Rosenfeld ([23]) sees the reluctance to accept
the Copenhagen view not so much in wavering bourgeois ten-
dencies than in Western Man's affliction with Aristotelian
logic, as against Eastern sages less prejudiced in favor of cold
Western logic and thus better equipped to 'understand Bohr'.
But when he rejects as unjustified the contention that the
Copenhagen doctrine is 'no longer concerned with things
but only with the way of speaking about things; science is de-
graded from a quest for truth to a verbal exercise', then he may
blame this condemnation of Copenhagen on his own words:
'While the great masters [Planck, Einstein, Born, Schrödinger]
were vainly trying to eliminate the contradictions in Aristotelian

fashion by reducing one aspect to another, Bohr realized the futility of such attempts. He knew that we have to live with this dilemma . . . and that the real problem was to refine the language of physics so as to provide room for the coexistence of the two conceptions.' In contrast, I submit that the real problem was to dispose of contradictions by the unitary particle mechanics of wavelike appearances (Duane) and then to reduce the quantum rules of mechanics to a non-quantal basis, as proposed in this book.

45. *New approach to the teaching of quantum mechanics*

The textbooks on quantum theory start with the presentation of a few characteristic experimental results with particular emphasis on the contrast between particle tracks in cloud and bubble chambers against maxima and minima of intensity in diffraction patterns. Thereafter, instead of simply concluding that *something periodic* is involved in those wavelike appearances (namely, the quantum rules connecting periodicities of bodies in space and time with their momentum and energy activity) the student is exposed to the wrong physics that each particle is, or manifests itself as, a wavicle, and to the evasive philosophy of dualism according to which it is meaningless to ask what is the real constitution of matter, and all the rest. After sufficient indoctrination the student is now prepared to accept the various rules of matrix and operator calculus together with the Schrödinger equation at face value. His lack of understanding is confirmed and even encouraged by a well-known authority, F. Dyson([24]), who writes approvingly: 'The student begins by learning the tricks of the trade. . . . Then he begins to worry because he does not understand what he is doing. This stage often lasts six months or longer, and it is strenuous and unpleasant. Then quite unexpectedly . . . the student says to himself: "I understand quantum mechanics", or rather he says: "I understand now that there isn't anything to be understood". . . . For each new generation of students there is less resistance to be broken down before they feel at home with quantum ideas.' We are not amused in spite of the fact that the last sentence cheerfully confirms the inheritance of acquired characteristics from one generation to another, even of such detrimental ones as breaking down to *not understanding* and yet 'feeling at home with quantum ideas'. In contrast, I think we owe it to the student to inform him from the beginning that quantum theory *can* be understood without linguistic artifices and dual manifestation; that the conservation laws of mechanics are valid also in the atomic domain; that there are special quantum rules of probability restricting their application; that the selection rules ($\Delta E = h\nu$, etc.) are *explainable* as consequences of regulative postulates of symmetry and invariance; and finally that the theory connects real data of instruments rather than mental states of observers. Altogether, expecting the student as a good quantum soldier never to ask 'why' may have its merits for those in a hurry to learn 'the tricks of the trade'. But I still think that a systematic approach starting from the ground up is not only intellectually satisfactory but time-saving. I here do not think of such highly mathematical and abstract books as those of von Neumann and Ludwig which of course were never meant for the beginner. A much shorter and yet systematic approach is typified by Margenau's analysis of the essentials of the quantum formalism which has led him to the following four axioms, quoted here without explanatory text:

Axiom I. To every physical observable corresponds a mathematical operator. In particular, to the observable $p =$ momentum belongs the differential operator $(h/2i\pi)\, \partial/\partial q$ where q is the space coordinate conjugate to p.

Axiom II. To every physical state S of a mechanical system corresponds a mathematical function $\psi_S(q)$. In particular, the state function belonging to a state of momentum value p is $\psi_p(q) = \exp(2i\pi pq/h)$, in consequence of the p-operator of Axiom I.

Axiom III. The only values which the measurement of a physical observable can yield are the eigenvalues of the corresponding operator.

Axiom IV. Let A be an observable with eigenvalues A_n belonging to the eigenfunctions $\psi_n(q)$. The *probability* of finding the value $A = A_n$ when the system in the state S is subjected to an A-measurement is $P_S(A_n) = |\psi_S(A_n)|^2$ where $\psi_S(A_n) = \int \psi_S(q) \cdot \psi_n(q) \cdot dq$.

Certainly there is nothing wrong with these four propositions. I only contend that, as an introduction, or as an effort to go down to essentials, they do not eliminate the *ad hoc* element from the foundations. Thus, the adept will ask *why* must he accept every observable to correspond to an 'operator'. And even the expert might like to know more elementary reasons why the observable p corresponds to the differential operator $(h/2i\pi)\, \partial/\partial q$. In fact, Axiom I contains most of the quantum theory in a nutshell, introduced by *fiat*, rather than derived from a non-quantal basis. And Axiom IV introduces the superposition of probability amplitudes, again by *fiat*; and the important two-way symmetry of the probabilities is not mentioned. Thus, when according to Margenau, 'an axiom is any proposition of physical or mathematical content which cannot be derived from other propositions', then his propositions I to IV do not belong to this class. Our counter-proposal is kept very brief because all details were developed in the foregoing chapters. It reads as follows:

Elementary quantum mechanics of matter has to do with material bodies composed of particles with coordinates and momenta, q and p, and with various observable quantities $A, B, C \ldots$ defined as functions of the q's and p's. Any observable quantity A can have many values, A_1 and A_2 and A_3 and so forth. Whether these characteristic or 'eigenvalues' form a continuity or a discrete set depends on the special qualities of the system and on the observable A under consideration. All this is trivial.

(*I*) It is a significant fact, however, that in the atomic domain there are 'observables' in the narrower sense of being connected with one another by definite probability relations. More specifically, when an A-measurement has yielded the value $A = A_k$, then subsequent B-measurements yield a *statistical distribution* over the values $B_1 B_2 \ldots$ with relative frequencies $P(A_k \to B_j)$. The probabilities P connecting various A- with various B-values may be arranged in a matrix (P_{AB}), the *rows* of which sum up to unity. For the sake of simplicity assume the P-matrices to be finite.

(*II*) In correspondence with the classical reversibility of mechanical processes, there is the theorem of a *two-way symmetry* of the probabilities, $P(A_k \to B_j) = P(B_j \to A_k)$, so that the arrows in this notation can be omitted. An immediate consequence of the P-symmetry is that not only the rows but also the columns of every P-matrix sum up to unity, and also that the rows and columns have the same number M of elements. In other words, the P-matrices are *unit sum squares* (doubly stochastic), and all observables A and B, etc., have *common multiplicity* M of their eigenvalues. (Actually, in many cases M is infinite.) Without P-symmetry there could not be any statistical equilibrium.

(*III*) We introduce the assumption or postulate that the various P-matrices are mutually *interdependent* so that the matrix (P_{AC}) is determined, or at least restricted in its elements, when (P_{AB}) and (P_{BC}) are given. The interdependence law is to be general, that is *symmetric* with respect to the various observables A, B, C, and also *transitive*, so as to hold as well for any other combination of observables A, B, D and B, C, D and so forth. There is but one possible interdependence theorem between unit sum squares satisfying the conditions of generality (symmetry and transitivity). It is the theorem of orthogonal transformation and its generalization, unitary transformation, the former with the help of real symmetric cosines, $\alpha(A_k \to B_j) = \alpha(B_j \to A_k)$ and $P = \alpha^2$, the latter with complex hermitian (symbolic) cosines

$\psi(A_k \to B_j) = \psi^*(B_j \to A_k)$ and $P = |\psi|^2$, with ψ as 'probability amplitudes'.

(*IV*) Whereas the quantity ψ also stands for the 'tensor unity' with $\psi_{kk'} = \delta_{kk'}$, an observable A represents a tensor in general, with $A_1 A_2 \ldots$ as its principal or 'eigenvalues', whereas the tensor components $A_{mn} = A_{nm}$ represent the mean values of A in transition from state m to n. A frequent problem is that of finding the eigenvalues when a set of transition values is given, by a 'transformation to principal axes'.

(*V*) Quantum Dynamics is to be invariant with respect to shifts of zero points in q- and p-space, as well as to shifts of zero points of energy and time, that is, Galileo and/or Lorentz invariance. In particular, any observable $S(q)$ shall have transition values $S_{pp'}$ depending on $p - p'$ only. And any observable $T(p)$ shall have transition values $T_{qq'}$ depending on $q - q'$ only. This requirement leads by mathematical necessity to the conclusion that ψ_{pq} must have the complex periodic form $\psi_{pq} = \exp(ipq/\text{const})$, whereby the constant, being of the dimension of an action, is usually denoted as $h/2\pi$. This concludes the derivation of quantum mechanics on the basis of simple non-quantal postulates. If the mathematical method of the quantum theory sometimes looks quite abstract—observable quantities are replaced by symbolic operators subjected to a non-commutative algorism involving Planck's constant—one may be assured that this mathematical language is but a shorthand notation describing relations which could as well be reduced to ordinary addition and multiplication of various data if one had enough paper, ink, and patience.

Summary

In contrast to the physical content of the previous chapters, the present one is an indictment of the current quantum philosophy of Bohr and Heisenberg and their followers. It rests on two main arguments:

(1) It is dubious metaphysics to give up the search for an

objective physical reality and to stop at two subjective pictures.

(2) It is dubious physics to ignore the third quantum rule which is the key to a unitary mechanical explanation of wavelike phenomena and disposes of the baroque hypothesis of double manifestation.

More specifically I object to a *quantum logic* which:

(3) sees a dualistic symmetry between a thing, a kickable particle, and a non-kickable symbolic wavelike curve representing probabilities of atomic events;

(4) replaces statistical distribution data which are uncertain to predict, by indeterminacy of their existence, and in particular declares that the momentum of an electron, though measurable from its deflection, is a meaningless or non-existent momentum, presenting this 'renunciation' as a philosophical discovery;

(5) counts only 'direct measurements' when *in fact* every measurement has indirect ingredients, only (q, t)-measurements, that is coincidences being direct;

(6) denies that anything happens between two measurements when *in fact* one can reconstruct what has happened between two direct (q, t)-measurements in the form of an exact path connecting them;

(7) presents the relations $E = h\nu$ and $p = h/\lambda$ as connections between two subjective pictures, when there are *objective quantum rules*, $\Delta E = h/\tau$ and $\Delta p = h/L$ which have nothing to do with 'pictures' but are plain physics and describe the activity of time- and space-periodic bodies (oscillators crystals, etc.);

(8) regards the Second Quantization as an argument for a dual nature of matter when *in fact* it is but a mathematical transformation of the theory of Rutherford's atomic model into an utterly complicated formal schema which could be augmented by dozens of other equally complicated formalisms;

(9) accepts as proof for the inverse proportionality of δE

and δt an argument which uses the direct proportionality of the same two quantities;

(10) considers as enlightening the story of a wave packet of knowledge or of absence of knowledge contracting with super-luminal velocity, when *in fact* this remainder of the defunct idea of matter waves has become a medley of inconsistencies resting on arbitrary changes of word meanings;

(11) believes that the consciousness of human observers is involved in an atomic measurement when there is no need of consciousness for an ordinary measurement;

(12) declares that the question of the real constitution of matter is meaningless and maintains that 'the real problem is to refine the language of physics', that we must not 'think in objects' any more, and give up 'insistence on the postulate of complete logical clarification'.

(13) indulges in 'physical and epistemological license . . . trying to satisfy positivistic, realistic, and idealistic doctrines all at once' (W. Yourgrau,([25])), the result being an uncontrolled fusion of contradictory ideas which no dialectics can reconcile.

When cleared of obsolete connotations and ideological ballast carried over from the 1920's, the quantum theory of matter emerges as a consistent unitary theory of matter particles and systems of particles, with statistical rules connecting their reactions to instruments of observation. Far from being irreducible axioms, the quantum principles can be derived from a simple non-quantal basis. It is this *demystification* of quantum mechanics, rather than the *myth-raking* of the last chapter, which constitutes the main object of this volume.

UNITARY TRANSFORMATION

We introduce the postulate that the P-matrices are mutually interdependent so that (P_{AB}) and $P(_{BC})$ either determine (P_{CA}), or at least restrict it, by a *general* interdependence theorem. When describing the required interdependence theorem as a functional relation $F[P_{AB}, P_{BC}, P_{CA}] = 0$, abbreviated $f(A, B, C) = 0$, *generality* is to mean *symmetry* in the letters A, B, C as well as *transitivity* of the functional relation, so that the three relations $f(A, B, C)$, $f(A, B, D)$, and $f(A, C, D) = 0$ yield, by elimination of A, the relation $f(B, C, D) = 0$. The elimination process itself runs as follows, using the letter ω instead of P:

(a) From $F[\omega_{AB}, \omega_{BC}, \omega_{CA}]$, abbreviated $f(A, B, C) = 0$ and $f(A, B, D) = 0$ eliminate ω_{AB} so as to leave a relation between $\omega_{BC}, \omega_{CA}, \omega_{BD}, \omega_{DA}$.

(b) Use it to express ω_{DA} in terms of $\omega_{BC}, \omega_{CA}, \omega_{BD}$ and substitute this expression for ω_{DA} in $f(A, C, D) = 0$.

(c) The result, which now contains the four letters A, B, C, D ought to reduce to $f(B, C, D) = 0$ in order to satisfy *transitivity*. The symbol ω may signify a numerical quantity, or a vector with several components, or a matrix with several rows and columns. In all cases the operation of addition and multiplication must be defined.

We now are going to try out systematically various functions f *symmetric* in A, B, C, first linear, then second, third, etc., order functions and test them for *transitivity* according to the procedure above.

(I) The only symmetric linear combination between two-index quantities named ω reads

$$f_1(A, B, C) = \omega_{AB} + \omega_{BC} + \omega_{CA} - k = 0.$$

Transitivity is satisfied only when the constant $k = 0$ and when

$$\omega_{AB} = -\omega_{BA}, \quad \omega_{AA} = 0, \quad \omega_{AB} + \omega_{BC} + \omega_{CA} = 0.$$

However, this triangular addition theorem cannot serve as a model for a general connection theorem between our P-matrices since $(P_{AA}) = (1)$ can never arise from $(\omega_{AA}) = (0)$.

(II) Next we try second-order function $f_2(A, B, C)$ of quantities (numbers, vectors, matrices) denoted as ϕ.

$(\phi_{AB} + \phi_{BC} + \phi_{CA})^2$ is identical with (I). $\phi_{AB}^2 + \phi_{BC}^2 + \phi_{CA}^2$ is transitive only under the impossible condition $\phi_{AB}^2 = -\phi_{BA}^2$.

There remains the symmetric form:

$f_2(A, B, C) = \phi_{AB}\phi_{BC} + \phi_{BC}\phi_{CA} + \phi_{CA}\phi_{AB} = 0$ which passes the transitivity test only when $\phi_{AB} = -\phi_{BA}$, hence $\phi_{AA} = 0$ which renders it unfit to serve as a model for $(P_{AA}) = (1)$.

(III) The simplest *symmetric* third-order connection between three two-index quantities called ψ is the product:

$$f_3(A, B, C) = \psi_{AB}\psi_{BC}\psi_{CA} - k = 0.$$

It becomes *transitive* for $k = 1$ and

$$\psi_{AB}\psi_{BA} = 1, \quad \psi_{AA} = 1, \quad \psi_{AB}\psi_{BC}\psi_{CA} = 1.$$

As a product theorem for matrices (ψ) it is fit to serve as a model for a general interdependence law between the P-matrices since $(P_{AA}) = (1)$ can rest on $(\psi_{AA}) = (1)$.

There are other third-order functions f_3 symmetric in A, B, C such as $f_3 = \psi_{AB}^2\psi_{BC} + \psi_{BC}^2\psi_{CA} + \psi_{CA}^2\psi_{AB} - k = 0$. However, since they contain ψ_{AB} in first as well as in second order, they lead, by the elimination process (a), (b), (c) above to bivalent *irrational* expressions and thus cannot pass the transitivity test. The same holds for functions f_4, f_5, etc., which inevitably lead to *multivalent irrational* or *transcendent results* in step (c) of the elimination. They cannot be transitive when ψ signifies a number, and even less when ψ signifies a matrix with rows and columns.

We thus arrive at the conclusion: The multiplication theorem

$$(\psi_{AB})\,(\psi_{BC}) = (\psi_{AC}) \text{ with } (\psi_{AB})\,(\psi_{BA}) = (1) \text{ and } (\psi_{AA}) = (1)$$

is the *only* one which connects triples of matrices (ψ_{AB}), (ψ_{BC}), and (ψ_{AC}) in a *univalent*, symmetric, and transitive fashion whereby (ψ_{AA}) is not the matrix (0). Notice that the product $(\psi_{AB})\,(\psi_{BA}) = (1)$ reduces to the sum rule, $\sum|\psi|^2 = 1$ for every row and every column of every matrix $(|\psi|^2)$, supposing that

$$\psi(A_k, B_j) = \psi^*(B_j, A_k) \text{ so that}$$

$$\psi(A_k, B_j) \cdot \psi(B_j, A_k) = |\psi(A_k, B_j)|^2 = |\psi(B_j, A_k)|^2.$$

The question is whether the matrix elements ψ of the formal schema (1) can be identified with the probabilities P. This is not the case since all P's are positive, whereas there must be positive as well as negative ψ's in order to yield vanishing non-diagonal elements in the product $(\psi_{AB})\,(\psi_{BA}) = (1)$. Yet there must be a close connection between ψ and P as may be seen from the following list of qualities possessed by ψ and required of P, the only difference being that the ψ-interdependence theorem is univalent (and is the only possible univalent one, see above) so that *a triangular P-interdependence law is necessarily multivalent.*

ψ-theorem P-law	(1) triangular (1) triangular	(2) symmetric (2) symmetric	(3) transitive (3) transitive
	both the same	same	same
ψ-theorem P-law	(4) $\sum|\psi|^2 = 1$ (4) $\sum P = 1$	(5) univalent (5) not unival	(6) $(\psi_{AA}) = (1)$ (6) $(P_{AA}) = (1)$
	same	different	same

And since every individual $\psi(A_k, B_j)$ as well as every individual $P(A_k, B_j)$ depends on A_k and B_j only, rather than on other indexed letters, the relation must be between the individuals $\psi(A_k, B_j)$ and $P(A_k, B_j)$, the first being a certain function of the second which, because of the generality of the P-law symmetric in all indexed letters, must be one *common* function F. Thus we arrive at the conclusion $P = F(\psi)$. Next, in order to satisfy all six conditions above for positive P's, the function can only be $P = |\psi|^2$. As foreseen, the P-law is not univalent: a given P determines only the absolute value of the corresponding complex ψ, but leaves its phase open. The result of our considerations is that unitary transformation via probability amplitudes ψ is the *only possible way* of connecting the positive two-way-symmetric quantities P by a general, that is symmetric and transitive, interdependence theorem. Unitary which includes orthogonal transformation, is not an oddity to be accepted at face value, or 'a consequence of the dualism of the wave picture and the particle picture'. It rather is the only conceivable way in which the probabilities can satisfy a general mutual interdependence theorem under the non-committal non-quantal postulate of law and order rather than chaos.

So far there is no reason for using complex quantities. ψ might be *real* in which case the asterisk * does not mean anything. The reason for complex quantities will be seen in Appendix 3.

TENSOR FORMALISM

If the probability metric of quantum theory is isomorphic with 'unitary transformation' in M-dimensional Hilbert space, the same then holds for other parts of quantum mechanics. In particular, observables play the part of tensors in the following manner. There are various sets of orthogonal axes, the set $a_1 a_2 \ldots a_M$, the set $b_1 b_2 \ldots b_M$, and so forth. They are characterized as 'orthogonal sets' by virtue of the 'generalized' cosine or ψ-relations

$$\psi_{aa'} = \delta_{aa'}, \quad \psi_{bb'} = \delta_{bb'}, \text{ etc.,} \tag{1}$$

together with the Hermitian condition $\psi_{ab} = \psi_{ba}^*$. The M axes a point in those directions in which a certain tensor A has its characteristic or principal or eigenvalues:

$$A_{aa'} = 0 \text{ for } a \neq a', \text{ but } A_{aa} = A_a \tag{2}$$

$$A_{bb'} = A_{b'b}^*, \text{ also } A_{bc} = A_{cb}^* \tag{3}$$

The complex quantities (3) are denoted as *transition values* or mean values of A in transition from state b to c (or vice versa), hence the eigenvalues A_a can be denoted as transition values of A from a state a to itself, for the following physical reason.

Suppose our atom, or mechanical system in general, is first ascertained to be in the state b, that is in one of the eigenstates of an observable B with eigenvalue B_b. It is now subjected to an A-meter test which may yield the result A_1 or A_2 etc., in the resulting final state a_1 or a_2 etc., arriving there with probabilities P_{ba}. The mean value of A obtained in a series of such tests, starting from state b then is $\sum_a P_{ba} \cdot A_a$. Summations are always taken over the complete set of orthogonal states, over $a_1 a_2 \ldots$ in the present case. Now, the latter expression is the definition of the mean value of A in the state b, or mean value of A in transition from state b to itself; it is denoted as

$$A_{bb} = \sum_a P_{ba} A_a = \sum_a \psi_{ba} A_a \psi_{ab}. \qquad (4)$$

Its natural generalization within the tensor calculus then is

$$A_{bc} = \sum_a \psi_{ba} A_a \psi_{ac} \qquad (5)$$

denoted as the mean value of A in transition from state b to c, or in short, *transition value* of A from b to c. The transition value A_{bc} is expressed here in terms of the eigenvalues of A. But A_{bc} can also be expressed in terms of another set of transition values A_{mn}, from the eigenstates of any observable M to those of any other observable N,

$$A_{bc} = \sum_m \sum_n \psi_{bm} A_{mn} \psi_{nc} \; (= \sum_k \sum_l \psi_{bk} A_{kl} \psi_{lc}). \qquad (6)$$

Again notice the last sentence of Appendix 1.

THE WAVE FUNCTION

The periodic form of the probability amplitude function together with the fact that complex unitary rather than real orthogonal transformation is basic for the quantum mechanics can be derived from the following postulate, or definition, of a conjugate pair of a linear coordinate q and momentum p: The transition values ($=$ tensor components) $T_{pp'}$ of every observable $T(q)$ are to depend on $p - p'$ only. And the transition values $S_{qq'}$ of every observable $S(p)$ are to depend on $q - q'$ only.

We begin with the general formula, valid for unitary as well as for the special case of orthogonal transformation, using integrals instead of sums:

$$T_{pp'} = \int \psi_{pq} \, T(q) \, \psi_{qp'} \, dq. \tag{1}$$

If the left-hand side is to depend on $p - p'$ for *every* function $T(q)$, this must also hold for the special case that $T(q)$ is a Delta function, $D(q) = \delta(q - q_0)$ vanishing everywhere except at one place $q = q_0$. This reduces the last equation to the simple product.

$$D_{pp'} = \psi_{pq_0} \, \psi_{q_0 p'}$$

for any chosen value q_0 for which we write q again. The question now is which function $\psi(p, q) = \psi_{pq}$ under the condition $\psi_{qp'} = \psi^*_{p'q}$ of unitary transformation (which still includes orthogonal transformation with real ψ), satisfies

$$\psi_{pq} \, \psi^*_{p'q} = f(p - p'). \tag{2}$$

In order to find out, expand both factors into power series in p:

$$\psi_{pq} = a_0 \left[1 + A_1 p + A_2 p^2 + \ldots \right], \tag{3}$$
$$\psi^*{}_{p'q} = a_0^* \cdot \left[1 + A_1^* p' + A_2^* p'^2 + \ldots \right].$$

The coefficients a_0 as well as $A_1 \, A_2 \ldots$ are functions of q alone. Take the product of the two series (3) and order it with respect to terms linear, quadratic, cubic, etc., in p and p'. In order that the product be a function of $p - p'$ it first turns out that we must have $A_1 = -A_1^*$, that is A_1 must be purely imaginary, $A_1(q) = i \cdot \alpha(q)$ where $\alpha(q)$ is a *real* function. Next, in order that the terms quadratic in p and p' will also depend on $p - p'$ only, p and p' must occur in the combination $p^2 - 2pp' + p'^2 = (p - p')^2$ which determines the ratio $A_2/A_1 = i\alpha/2$. Continuing in this fashion the coefficiencts A_n turn out to be such that the series (3) simply represents the exponential function

$$\psi_{pq} = a_0(q) \cdot \exp[i \cdot \alpha(q) \cdot p] \tag{4}$$

with real function $\alpha(q)$ in the exponent. The complex form of ψ_{pq} excludes orthogonal in favor of complex unitary transformation. Now exchange the letters q and p and apply the same considerations; this leads to

$$\psi_{qp} = b_0(p) \exp[i \cdot \beta(p) \cdot q] = \psi^*{}_{pq}$$

where β is a real function of p, hence

$$\psi_{pq} = b_0^*(p) \exp[-i \cdot \beta(p) \cdot q]. \tag{4'}$$

Comparison of (4) and (4') yields

$a_0(q) = b_0^*(p) = $ const., independent of both q and p,
$\alpha(q)/q = -\beta(p)/p = $ real constant,

so that finally

$$\psi_{pq} = \text{const} \cdot \exp[ipq \cdot \text{real const}].$$

The value of the real constant, usually written as $2\pi/h$, in terms of conventional units can, of course, not be obtained from any theoretical considerations. With

$$\psi_{pq} = \text{const} \cdot \exp(2i\pi pq/h) = \psi^*_{qp} \qquad (5)$$

any function $T(q)$ substituted on the right of (1) yields $T_{pp'}$ dependent on $p - p'$ only, and $S_{qq'}$ for any function $S(q)$ dependent on $q - q'$ only.

MASS SPECTRUM

The following considerations were proposed by the author together with L. H. Thomas as a first oversimplified approach toward obtaining a mass spectrum of elementary particles with no other basic assumption than that they all are characterized by one common universal length a or 'radius'. If mechanics, classical as well as relativistic, deterministic or quantal, gives equal rights to length and momentum, it follows immediately that a universal length a must be accompanied by a preferential momentum b. When b is expressed as mc with c as the universal velocity of light, one arrives at a preferential mass m. In order to use the mechanical symmetry of position (vector r) and momentum (vector p) we first introduce dimensionless quantities:

$$R = r/a \quad \text{and} \quad P = p/mc \tag{1}$$

belonging to dimensionless volume elements in space and momentum space

$$dV = dv/a^3 \quad \text{and} \quad dW = dw/(mc)^3. \tag{2}$$

Now, coming closer to quantum theory, irrespective of the physical meaning of symbols ψ and χ one has the following formulas of Fourier analysis:

$$\psi(R) = (\mu/2\pi)^{3/2} \iiint \chi(P) \cdot \exp[i\mu(P \cdot R)] \, dW, \quad \text{and}$$
its inversion $\hspace{9cm}$ (3)
$$\chi(P) = (\mu/2\pi)^{3/2} \iiint \psi(R) \cdot \exp[-i\mu(P \cdot R)] \, dV, \tag{4}$$

valid for any function ψ and its Fourier transform χ as well as for any value of the numerical parameter μ. If ψ and χ are to be probability amplitudes of quantum mechanics then μ would be a reduced rest mass

$$\mu = m \cdot 2\pi \, ca/h, \qquad (5)$$

whereby m, like μ, could have any value whatsoever.

The situation becomes different, however, when the volume elements dV and dW are replaced by the relativistic volume element dW_0 and a hypothetical dV_0 where

$$dV_0 = dV/\sqrt{R^2 + 1} \quad \text{and} \quad dW_0 = dW/\sqrt{P^2 + 1}. \quad (6)$$

Equations (3) (4) thus modified are inversions of one another only for certain selected *eigenfunctions*, $\psi = \chi^*$, and for corresponding eigenvalues of the parameter μ. A mathematical investigation by L. H. Thomas and the author† has shown that the eigenvalues are solutions of the transcendental equations ($\mathcal{J}_0$ and Y_0 are Bessel functions):

$$[\mathcal{J}_0(\mu)]^2 \cdot 2\pi\mu = 1 \quad \text{and} \qquad (7)$$
$$[Y_0(\mu)]^2 \cdot 2\pi\mu = 1. \qquad (8)$$

They are solved by the following μ-spectra

$$\mu = 0.16, \quad 1.87, \quad 2.92, \quad \ldots \qquad (7')$$
$$\mu = 0.03, \quad 1.38, \quad 3.42, \quad \ldots \qquad (8')$$

Since the magnitude of the length a is unknown, the μ-values yield only the *ratios* of the rest masses, $m = \mu h/2\pi ac$. Significant is the large ratio of the two first masses obtained from the Y_0-series, a ratio of about 1:46, followed by larger masses of about equal differences. Of course, one cannot expect that this speculation has a direct relation to reality. It only intends to show that there is a simple and primitive way of obtaining a mass spectrum of some sort from the hypothesis of a basic length a and a corresponding momentum, $b=mc$. Actually, the problem of the mass ratio of the elementary particles is vastly more complicated. In particular, what is observed as the rest mass may be heavily masked by other effects. And particles thought to be elementary might actually be composite. It would be presumptuous therefore to come forward with a so-called world formula for mass ratios at the present time.

†— A. Landé and L. H. Thomas, *Journal Franklin Inst.* 231, 63, 1941.

REFERENCES

CHAPTER I

1. W. Duane, *Proc. Nat. Acad. Sci., Wash.*, **9**, 158, 1923.
2. C. F. von Weizsäcker, *The World View of Physics*, University of Chicago Press, 1952.
3. P. Ehrenfest and P. Epstein, *Proc. Nat. Acad. Sci., Wash.*, **10**, 133, 1924 and **13**, 400, 1927.
4. B. Hoffman, *The Strange Story of the Quantum*, Dover Book, New York, 1959.
5. Sir James Jeans, *Nature*, 8 September 1934.
6. E. Schrödinger, *What is Life*, Cambridge University Press, 1944.

CHAPTER II

1. W. Heisenberg, *Physics and Philosophy*, Harper Torch Book, 1958.
2. R. D. Bradley, *Brit. J. Philos. Sci.*, **13**, 193, 1962; *Mind*, **68**, 193, 1959.
3. M. Black, in *Determinism and Freedom*, ed. by Sidney Hook, New York University Press, 1958.
4. M. Bunge, *Causality*, Harvard University Press, 1959.
5. P. Bridgman, *Sci. Mon.*, **79**, 32, 1954.
6. K. R. Popper, *The Logic of Scientific Discovery*, Hutchinson, London, 1956.
7. N. Bohr, in *Albert Einstein, Philosopher-Scientist*, ed. P. A. Schilpp, Evanston, Illinois, 1949.
8. J. von Neumann, *Mathematische Grundlagen der Quantenmechanik*, Springer, Berlin, 1932.
9. L. de Broglie, *La Physique Quantique, restera-t-elle indéterministe?*, Gauthier-Villars, Paris, 1953.
10. P. Feyerabend, *Z. Phys.*, **148**, 551, 1957; G. Schultz, *Ann. Phys.*, **3**, 94, 1959.
11. Chs. S. Peirce, quoted from *Readings in the Philosophy of Science*, ed. Philip P. Wiener, New York, 1953.
12. K. R. Popper, *Brit. J. Philos. Sci.*, **1**, 1, 1950.
13. *Determinism and Freedom*, ed. Sidney Hook, New York University Press, 1958.

CHAPTER III

1. *Leibnitz Selections*, ed. by Philip P. Wiener, Scribners, New York, 1951.
2. H. Reichenbach, *Philosophical Foundations of Quantum Mechanics*, University of California Press, 1944.
3. V. Hinshaw, *Philos. Sci.*, **26**, 310, 1959

CHAPTER IV

1. A. Landé, *Foundations of Quantum Theory*, Yale University Press, 1955.
2. N. Bohr, *Dialectica*, **2**, 312, 1948.

REFERENCES

CHAPTER V

1. A. Landé, *Foundations of Quantum Theory*, Yale University Press, 1955.
2. E. Schrödinger, *Statistical Thermodynamics*, Cambridge University Press, 1946.
3. J. von Neumann, *Mathematische Grundlagen der Quantenmechanik*, Springer, Berlin, 1932.
4. L. Rosenfeld, *Science Progress*, **163**, 393, 1953.
5. L. Rosenfeld, *Nuclear Physics*, **1**, 139, 1956.
6. W. Heisenberg, *Physics and Philosophy*. Springer, Berlin, 1932.
7. H. Reichenbach, *The Direction of Time*, University of California Press, 1956.

CHAPTER VI

1. N. Bohr, *Atomic Physics and Human Knowledge*, John Wiley, New York, 1958.
2. P. Bridgman, *Sci. Mon.*, **79**, 32, 1954.
3. W. Heisenberg in *Niels Bohr and the Development of Physics*, Pergamon Press, Oxford, 1955.
4. P. Jordan in *The Axiomatic Method*, North-Holland Publ. Co., 1959.
5. J. Lennard-Jones, *Sci. Mon.*, **80**, 175, 1955.
6. W. Pauli, in *Niels Bohr and the Development of Physics*, Pergamon Press, Oxford, 1955.

CHAPTER VII

1. C. F. von Weizsäcker, *The World View of Physics*, University of Chicago Press, 1952.
2. F. Bopp, *Heisenberg Festschrift*, Vieweg, Braunschweig, 1961.
3. P. P. Ewald, *Kristalle und Röntgenstrahlen*, Springer, Berlin, 1923.

CHAPTER VIII

1. C. F. von Weizsäcker, *The World View of Physics*, University of Chicago Press, 1952.
2. W. Heisenberg, *The Physical Principles of Quantum Theory*, Chicago University Press, 1930.
3. Sir James Jeans, *Nature*, 8 September 1934.
4. Sir Charles Darwin, in *Observation and Interpretation*, ed. S. Körner, Butterworths, 1957.
5. M. Born, *Philos. Quarterly*, **3**, 139, 1953.
6. N. Bohr, *Dialectica*, **2**, 312, 1948.
7. W. Heisenberg in *Niels Bohr and the Development of Physics*, Pergamon Press, Oxford, 1955.
8. W. Heitler, *The Quantum Theory of Radiation*, Oxford University Press, 1936.
9. A. Einstein in *Albert Einstein, Philosopher-Scientist* (p. 672).
10. N. Bohr in *Albert Einstein, Philosopher-Scientist*.
11. R. Oppenheimer, *Science and the Common Understanding*, New York, 1953.
12. R. D. Bradley, *Brit. J. Philos. Sci.*, **13**, 193, 1962.
13. K. R. Popper, *The Logic of Scientific Discovery*, Hutchinson, London, 1956.
14. M. Bunge, *Brit. J. Philos. Sci.*, **6**, 1 and 141, 1955.
15. P. Feyerabend, *Z. Phys.*, **148**, 551, 1957.
16. H. Margenau, *The Nature of Physical Reality*, McGraw-Hill, New York, 1950.

REFERENCES

17. H. Reichenbach, *The Philosophical Foundations of Quantum Mechanics*, University of California Press, 1944.
18. H. Margenau, *Philos. Sci.*, **25**, 23, 1958.
19. W. Heisenberg, *Physics and Philosophy*, Harper Torchbook, New York, 1962.
20. Jean-Paul Sartre, *L'Etre et le Néant*, Paris, 1943.
21. L. Rosenfeld, *Science Progress*, **163**, 393, 1953.
22. Chr. Caudwell, *The Crisis in Physics*, Bodley Head, London, 1939.
23. L. Rosenfeld, *Physics Today*, October 1963.
24. F. Dyson, *Scient. Amer.*, September 1958.
25. W. Yourgrau, 'On the Reality of Elementary Particles', in *The Critical Approach to Science and Philosophy*, Macmillan, New York, 1964.

INDEX

The main references are in italics

169

Printed in the United States
by Bookmasters

Printed in the United States
By Bookmasters